Autodesk Inventor 2020 Basics Tutorial

Tutorial Books

Download Resource Files from:

www.tutorialbook.info

Contents

Contents

Contents

Contents

Contents

Contents

Contents

Contents

INTRODUCTION

Autodesk Inventor as a topic of learning is vast and has a broad scope. It is a package of many modules delivering exceptional value to enterprises. It offers a set of tools, which are easy-to-use to design, document, and simulate 3D models. Using this software, you can speed up the design process and reduce product development costs.

This book provides a step-by-step approach for users to learn Autodesk Inventor. It is aimed for those with no previous experience with Inventor. However, users of previous versions of Inventor may also find this book useful for them to learn the new enhancements. The user will be guided from starting an Autodesk Inventor 2020 session to creating parts, assemblies, and drawings. Each chapter has components explained with the help of real-world models.

Scope of this book

This book is written for students and engineers who are interested to learn Autodesk Inventor 2020 for designing mechanical components and assemblies, and then create drawings.

This book provides a step-by-step approach for learning Autodesk Inventor 2020. The topics include Getting Started with Autodesk Inventor 2020, Basic Part Modeling, Creating Assemblies, Creating Drawings, Additional Modeling Tools, Sheet Metal Modeling, Assembly Tools, Dimensions and Annotations, and Model-Based Dimensioning.

Chapter 1 introduces Autodesk Inventor. The user interface and terminology are discussed in this chapter.

Chapter 2 takes you through the creation of your first Inventor model. You create simple parts.

Chapter 3 teaches you to create assemblies. It explains the Top-down and Bottom-up approaches for designing an assembly. You create an assembly using the Bottom-up approach.

Chapter 4 teaches you to create drawings of the models created in the earlier chapters. You will also learn to place exploded views, and part list of an assembly.

Chapter 5: In this chapter, you will learn the sketching tools.

Chapter 6: In this chapter, you will learn additional modeling tools to create complex models.

Chapter 7 introduces you to Sheet Metal modeling. You will create a sheet metal part using the tools available in the Sheet Metal environment.

Chapter 8 teaches you to create Top-down assemblies. It also introduces you to create mechanisms by applying joints between the parts.

Chapter 9: teaches you to apply dimensions and annotations to a 2D drawing.

Chapter 10: teaches you to add 3D annotations and tolerances to a 3D model.

Chapter 1: Getting Started with Autodesk Inventor 2020

This tutorial book brings in the most commonly used features of the Autodesk Inventor.

In this chapter, you will:

- Understand the Inventor terminology
- Start a new file
- Understand the User Interface
- Understand different environments in Inventor

In this chapter, you will learn some of the most commonly used features of Autodesk Inventor. Also, you will learn about the user interface.

In Autodesk Inventor, you create 3D parts and use them to create 2D drawings and 3D assemblies.

Inventor is Feature Based. Features are shapes that are combined to build a part. You can modify these shapes individually.

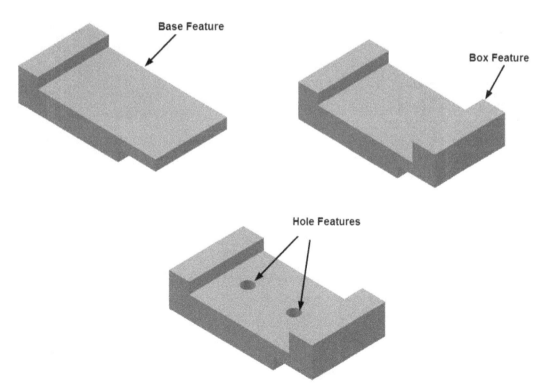

Most of the features are sketch-based. A sketch is a 2D profile and can be extruded, revolved, or swept along a path to create features.

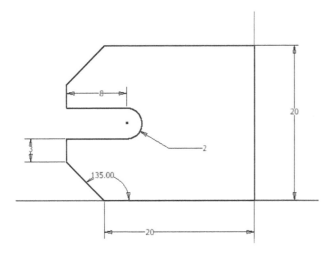
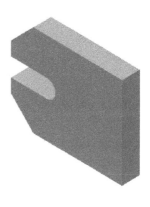

Inventor is parametric. You can specify standard parameters between the elements. Changing these parameters changes the size and shape of the part. For example, see the design of the body of a flange before and after modifying the parameters of its features.

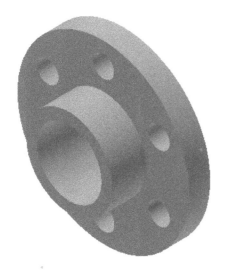
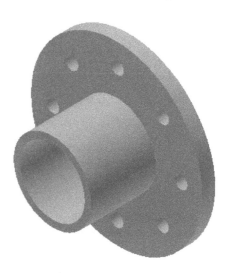

Starting Autodesk Inventor

- Click the Windows icon on the taskbar.
- Click **A** > **Autodesk Inventor 2020** > **Autodesk Inventor 2020**.
- On the ribbon, click **Get Started** > **Launch** > **New**.
- On the **Create New File** dialog, click the **Templates** folder located at the top left corner. You can also select the **Metric** folder to view various metric templates.
- In the **Part – Create 2D and 3D objects** section, click the **Standard.ipt** icon.
- Click **Create** to start a new part file.

Notice these essential features of the Inventor window.

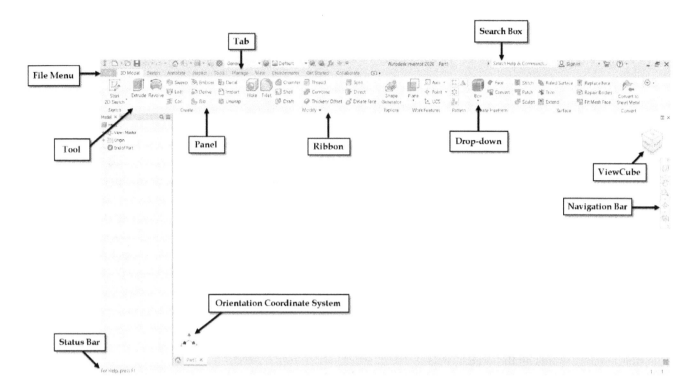

User Interface

Various components of the user interface are discussed next.

Ribbon

The ribbon is located at the top of the window. It consists of various tabs. When you click on a tab, a set of tools appear. These tools are arranged in panels. You can select the required tool from this panel. The following sections explain the various tabs of the ribbon available in Autodesk Inventor.

The Get Started ribbon tab

This ribbon tab contains tools such as **New, Open, Projects,** and so on.

The 3D Model ribbon tab

This ribbon tab contains the tools to create 3D features, planes, surfaces, and so on.

The View ribbon tab
This ribbon tab contains the tools to modify the display of the model and user interface.

The Inspect ribbon tab
This ribbon tab has tools to measure the objects. It also has analysis tools to analyze the draft, curvature, surface, and so on.

Sketch ribbon tab
This ribbon tab contains all the sketch tools.

Assemble ribbon tab
This ribbon tab contains the tools to create an assembly. It is available in an assembly file.

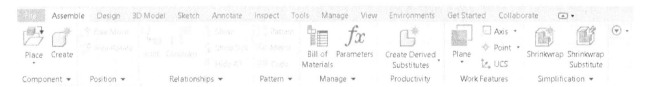

Presentation ribbon tab
This tab contains the tools to create the exploded views of an assembly. It also has the tools to create presentations, assembly instructions, and animation of an assembly.

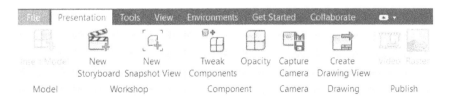

Drawing Environment ribbon tabs
In the Drawing Environment, you can create print-ready drawings of a 3D model. The ribbon tabs in this environment contain tools to create 2D drawings.

The Place Views ribbon tab

This ribbon tab has commands and options to create and modify drawing views on the drawing sheet.

The Annotate ribbon tab

This ribbon tab has commands and options to add dimensions and annotations to the drawing views.

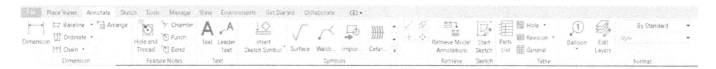

The Sheet Metal ribbon tab

The tools in this tab are used to create sheet metal components.

File Menu

This appears when you click on the **File** tab located at the top left corner. This menu contains the options to open, print, export, manage, save, and close a file.

Quick Access Toolbar

This is available at the top left of the window. It contains tools such as **New**, **Save**, **Open,** and so on.

You can customize this toolbar by clicking the down arrow at the right side of this toolbar.

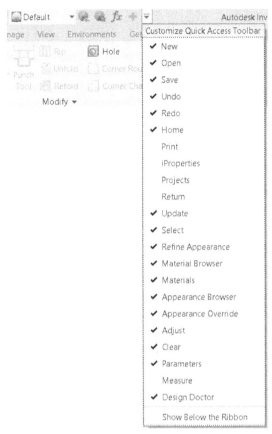

Browser window

This is located on the left side of the window. It contains the list of operations carried in an Autodesk Inventor file.

Status bar

This is available below the Browser window. It displays the prompts and the actions taken while using the tools.

Navigation Bar

This is located on the right side of the window. It contains the tools to zoom, rotate, pan, or look at the face of the model.

View Cube

It is located at the top right corner of the graphics window. It is used to set the view orientation of the model.

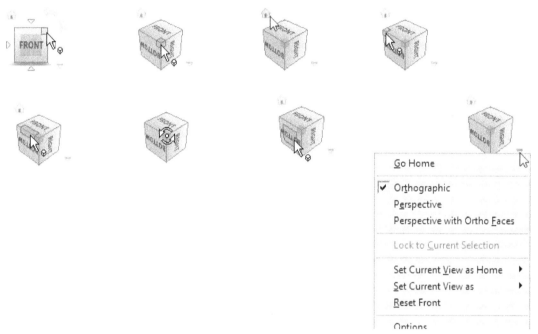

Shortcut Menus and Marking Menus

When you click the right mouse button, a shortcut menu along with a marking menu appears. A shortcut menu contains a list of some relevant options. The marking menu contains essential tools. It allows you to access the tools quickly. You can customize the marking menu (add or remove tools).

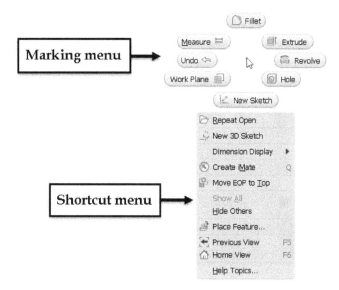

Dialogs

When you activate any tool in Autodesk Inventor, the dialog related to it appears. It consists of various options, which help you to complete the operation. The following figure shows the components of the dialog.

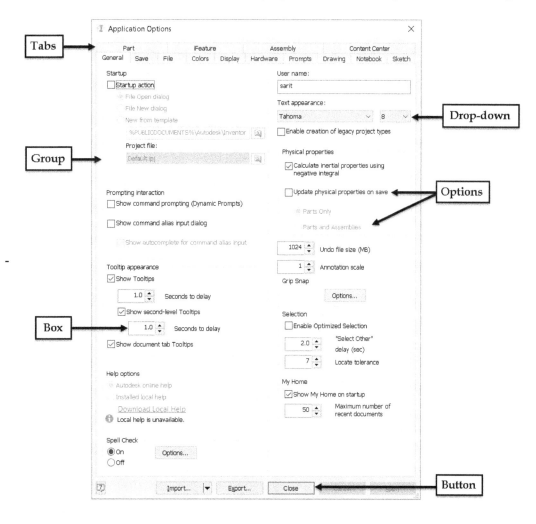

Customizing the Ribbon, Shortcut Keys, and Marking Menus

To customize the ribbon, shortcut keys, or marking menu, click **Tools > Options > Customize** on the ribbon. On the **Customize** dialog, use the tabs to customize the ribbon or marking menu, or shortcut keys.

For example, to add a command to the ribbon, select the command from the list on the left side of the dialog and click the **Add** >> button. If you want to remove a command from the ribbon, then select it from the right-side list and click the **Remove** << button. Click **OK** to make the changes to effect.

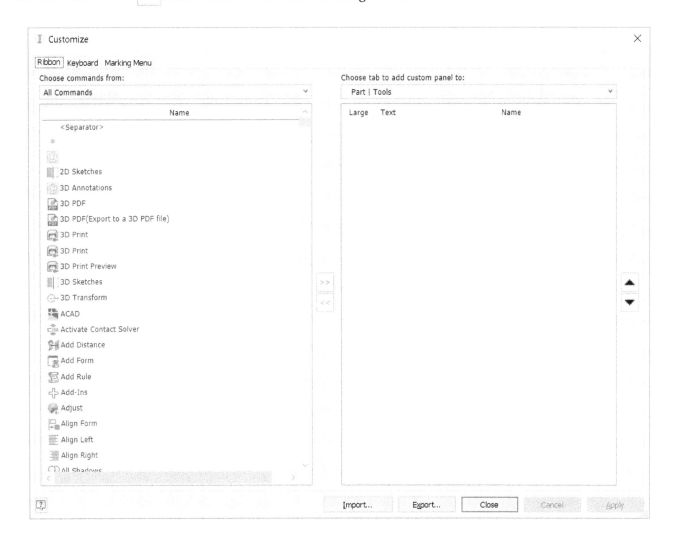

To add or remove panels from the ribbon, click the **Show Panels** icon located at the right side of the ribbon and check/uncheck the options on the fly out.

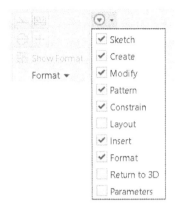

Color Settings

To change the background color of the window, click **Tools > Options > Application Options** on the ribbon. On the **Application Options** dialog, click the **Colors** tab on the dialog. Set the **Background** value to **1 Color** to change the background to plain. Select the required color scheme from the **Color Scheme** group. Click **OK**.

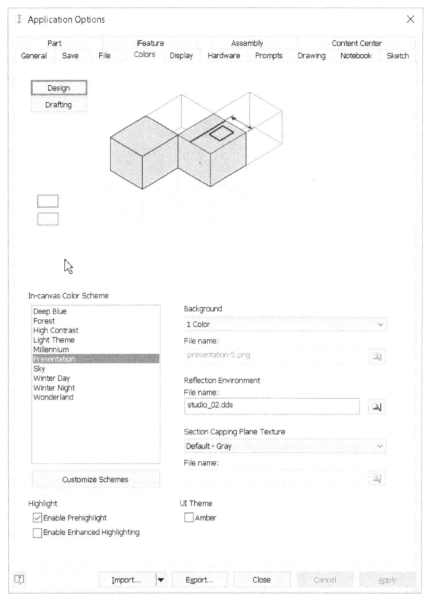

If you want to customize a scheme, click the **Customize Schemes** button; the **Color Scheme Editor** dialog appears. On this dialog, select the scheme to customize from the **Color Schemes** list. Next, expand the **Graphical Elements** tree and then click on the color swatch located next to the element of which the color is to be changed. Select a color from the **Color** dialog, and click **OK**. On the **Color Scheme Editor** dialog, drag the **Translucency** dragger to change the intensity of the color. Next click **OK** on the **Color Scheme Editor** and **Application Options** dialogs.

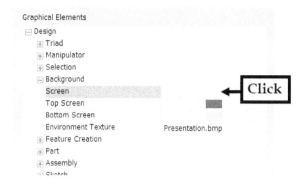

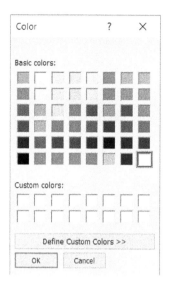

Chapter 2: Part Modeling Basics

This chapter takes you through the creation of your first Inventor model. You create simple parts:

In this chapter, you will:

- Create Sketches
- Create a base feature
- Add another feature to it
- Create revolved features
- Create cylindrical features
- Create box features
- Apply draft

TUTORIAL 1

This tutorial takes you through the creation of your first Inventor model. You will create the Disc of an Oldham coupling:

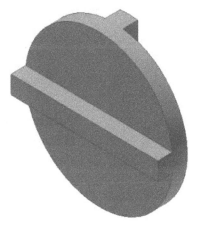

Creating a New Project

1. Start **Autodesk Inventor 2020** by double-clicking the **Autodesk Inventor 2020** icon on your desktop.
2. To create a new project, click **Get Started > Launch > Projects** on the ribbon.

3. Click the **New** button on the **Projects** dialog.
4. On the **Inventor project wizard** dialog, select **New Single User Project** and click the **Next** button.
5. Enter **Oldham Coupling** in the **Name** field.
6. Enter **C:\Users\Username\Documents\Inventor\Oldham Coupling** in the **Project(Workspace) Folder** box and click **Next**.
7. Click **Finish**.
8. Click **OK** on the **Inventor Project Editor** dialog.

9. Click **Done**.

Starting a New Part File

1. To start a new part file, click **Get Started > Launch > New** on the ribbon.
2. On the **Create New File** dialog, click the **Templates** folder located the top right corner.
3. Click the **Standard.ipt** icon located under the **Part – Create 2D and 3D Objects** section.
4. Click the **Create** button on the **Create New File** dialog.

A new model window appears.

Starting a Sketch

1. To start a new sketch, click **3D Model > Sketch > Start 2D Sketch** on the ribbon.

2. Click on the **XY Plane**. The sketch starts.

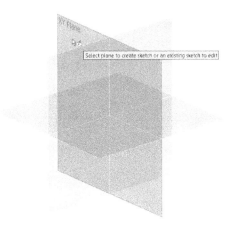

The first feature is an extruded feature from a sketched circular profile. You will begin by sketching the circle.

3. On the ribbon click **Sketch > Create > Circle > Circle Center Point**.

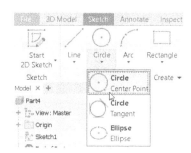

4. Move the cursor to the sketch origin located at the center of the graphics window, and then click on it.
5. Drag the cursor up to a random location, and then click to create a circle.

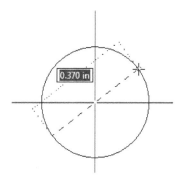

6. Press **ESC** to deactivate the tool.

Adding Dimensions

In this section, you will specify the size of the sketched circle by adding dimensions. As you add dimensions, the sketch can attain any one of the following states:

Fully Constrained sketch: In a fully constrained sketch, the positions of all the entities are fully described by dimensions, constraints, or both. In a fully constrained sketch, all the entities are a dark blue color.

Under Constrained sketch: Additional dimensions, constraints, or both are needed to specify the geometry completely. In this state, you can drag under constrained sketch entities to modify the sketch. An under constrained sketch entity is in black color.

If you add any more dimensions to a fully constrained sketch, a message box will appear showing that dimension over constraints the sketch. In addition, it prompts you to convert the dimension into a driven dimension. Click **Accept** to convert the unwanted dimension into a driven dimension.

1. On the ribbon, click **Sketch > Constrain > Dimension**.

2. Select the circle and click; the **Edit Dimension** box appears.
3. Enter **4** in the **Edit Dimension** box and click the green check.
4. Press **Esc** to deactivate the **Dimension** tool.

You can also create dimensions while creating sketch objects. To do this, enter the dimension values in the boxes displayed while sketching.

5. To display the entire circle at full size and to center it in the graphics area, use one of the following methods:

   - Click **Zoom All** on the **Navigate Bar**.
   - Click **View > Navigate > Zoom All** on the ribbon.

6. Click **Finish Sketch** on the **Exit** panel.

7. Again, click **Zoom All** on the **Navigate Bar.**

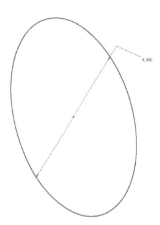

Creating the Base Feature

The first feature in any part is called a base feature. You now create this feature by extruding the sketched circle.

1. On the ribbon, click **3D Model > Create > Extrude.**

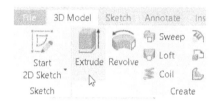

2. Type-in 0.4 in the **Distance A** box available on the **Extrude** Properties panel.
3. Click the **Direction > Default** icon under the **Behavior** section on the **Extrude** Properties panel.
4. Click **OK** on the **Extrude** Properties panel to create the extrusion.

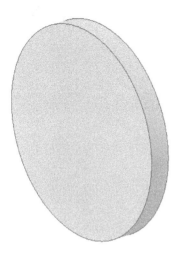

Notice the new feature, **Extrusion 1**, in the **Browser window**.

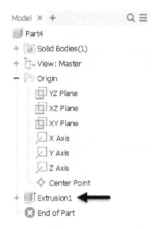

To magnify a model in the graphics area, you can use the zoom tools available on the **Zoom** drop-down in the **Navigate** panel of the **View** tab.

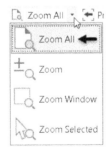

Click **Zoom All** to display the part full size in the current window.

Click **Zoom Window**, and then drag the pointer to create a rectangle; the area in the rectangle zooms to fill the window.

Click **Zoom**, and then drag the pointer. Dragging up zooms out; dragging down zooms in.

Click on a vertex, an edge, or a feature, and then click **Zoom Selected**; the selected item zooms to fill the window.

To display the part in different rendering modes, select the options in the **Visual Style** drop-down on the **Appearance** panel of the **View** tab.

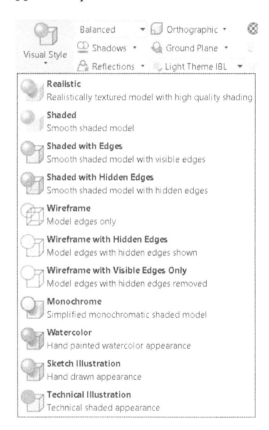

Realistic

Shaded

Shaded with Edges

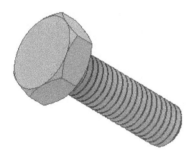

Shaded with Hidden Edges

Wireframe

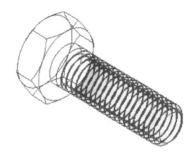

Wireframe with Hidden Edges

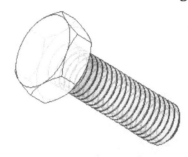

Wireframe with Visible Edges Only

Monochrome

Watercolor

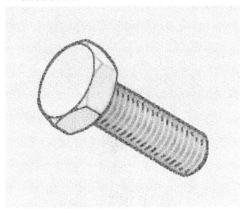

Sketch Illustration

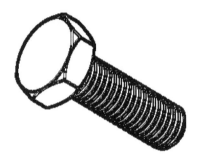

Technical Illustration

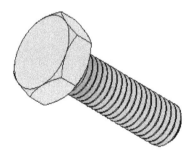

The default display mode for parts and assemblies is **Shaded with Edges**. You may change the display mode whenever you want.

Adding an Extruded Feature

To create additional features on the part, you need to draw sketches on the model faces or planes, and then extrude them.

1. On the ribbon, click **View > Appearance > Visual Style > Wireframe**.

2. On the ribbon, click **3D Model > Sketch > Start 2D Sketch**.
3. Click on the front face of the part.
4. Click **Line** on the **Create** panel.

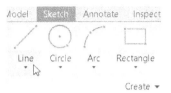

5. Click on the circular edge to specify the first point of the line.

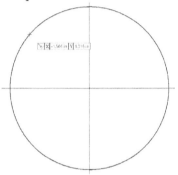

6. Move the cursor towards the right.
7. Click on the other side of the circular edge; a line is drawn.

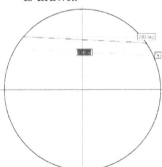

8. Draw another line below the previous line.

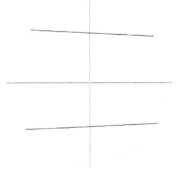

9. On the ribbon, click **Sketch > Constrain > Horizontal Constraint** .
10. Select the two lines to make them horizontal.
11. On the ribbon, click **Sketch > Constrain > Equal** .
12. Select the two horizontal lines to make them equal.
13. Click **Dimension** on the **Constrain** panel of the **Sketch** ribbon tab.
14. Select the two horizontal lines.
15. Move the cursor toward the right and click to locate the dimension; the **Edit Dimension** box appears.
16. Enter **0.472** in the **Edit Dimension** box and click the green check.

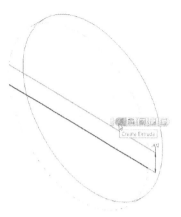

19. Click on the region bounded by the two horizontal lines.

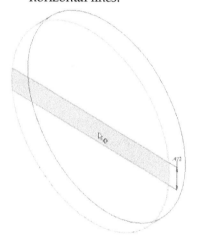

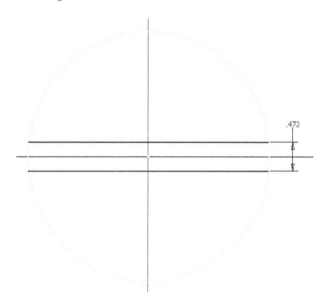

17. Click **Finish Sketch** on the **Exit** panel.
18. Click on the sketch, and then click **Create Extrude** on the **Mini Toolbar**; the **Extrude** Properties panel appears.

20. Enter **0.4** in the **Distance A** box on the **Extrude** Properties panel.
21. On the **Extrude** Properties panel, click the **Default** icon, and then **OK** to create the extrusion.

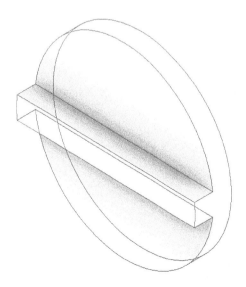

Part Modeling Basics

By default, the ambient shadows are displayed on the model. However, you can turn off the ambient shadows by clicking **View** tab > **Appearance** panel > **Shadows** drop-down, and then unchecking the **Ambient Shadows** option. The **Shadows** drop-down has two more options, which you use based on your requirement.

You can reuse the sketch of an already existing feature. To do this, expand the feature in the Browser Window, right click on the sketch, and select Share Sketch from the shortcut menu. You will notice that the sketch is visible in the graphics window. You can also unshare the sketch by right-clicking on it and selecting Unshare.

Adding another Extruded Feature

1. Click **Start 2D Sketch** on the **Sketch** panel of the **3D Model** ribbon tab.
2. Use the **Free Orbit** button from the **Navigate Bar** to rotate the model such that the back face of the part is visible.
3. Right click and select **OK**.
4. Click on the back face of the part.
5. Click **Line** on the **Create** panel.
6. Draw two lines, as shown below (refer to the **Adding an Extruded Feature** section to know how to draw lines). Make sure that the endpoints of the lines are coincident with the circular edge. Follow the next two steps, if they are not coincident.

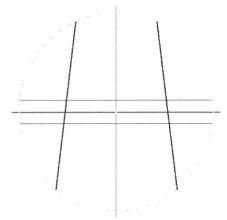

7. On the ribbon, click **Sketch > Constrain > Coincident Constraint**. Next, select the endpoint of the line and the circular edge.

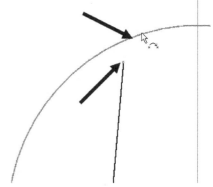

8. Likewise, make the other endpoints of the lines coincident with the circular edge.

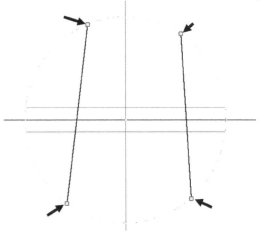

Skip the above two steps if the endpoints of the lines are coincident with the circular edge.

You can specify a point using various point snap options. To do this, activate a sketching tool, right click, and select Point Snaps; a list of point snaps appears. Now, you can select only the specified point snap.

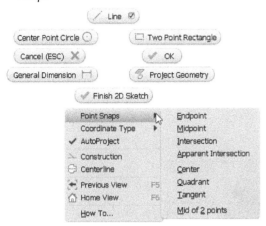

9. On the ribbon, click **Sketch > Constrain > Vertical Constraint** .

10. Select the two lines to make them vertical.

11. On the ribbon, click **Sketch > Constrain > Equal** .

12. Select the two vertical lines to make them equal.

13. Create a dimension of 0.472in between the vertical lines.

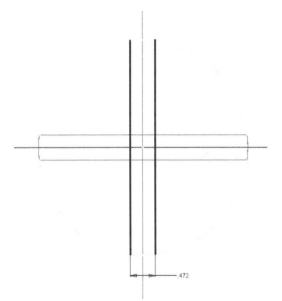

14. Click **Finish Sketch**.

15. On the ribbon, click **3D Model > Create > Extrude**.

16. Click inside the region enclosed by two lines, if they are not already selected.

17. Type 0.4 in the **Distance A** box on the **Extrude** Properties panel and click **OK**.

    To move the part view, click **Pan** on **Navigate Bar**, and then drag the part to move it in the graphics area.

18. On the ribbon, click **View > Appearance > Visual Style > Shaded with Edges**.

19. On the ribbon, click **View > Navigate > Home View** .

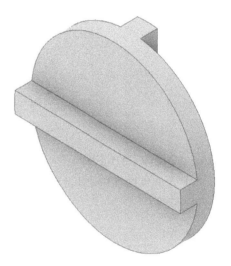

Saving the Part

1. Click **Save** on the **Quick Access Toolbar**.
2. On the **Save As** dialog, type-in **Disc** in the **File name** box.
3. Click **Save** to save the file.
4. Click **File Menu > Close**.

Note:

*.ipt is the file extension for all the files that you create in the Part environment of Autodesk Inventor.

TUTORIAL 2

In this tutorial, you create a flange by performing the following:

- Creating a revolved feature
- Creating cut features
- Adding fillets

Starting a New Part File

1. To start a new part file, click the **Part** icon on the Home screen.

Sketching a Revolve Profile

You create the base feature of the flange by revolving a profile around a centerline.

1. Click **3D Model > Sketch > Start 2D Sketch** on the ribbon.
2. Select the YZ plane.

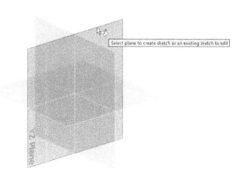

3. Click **Line** ╱ on the **Create** panel.
4. Create a sketch similar to that shown in the figure.

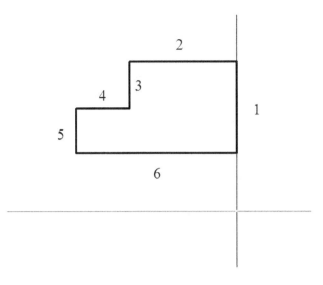

5. On the ribbon, click **Sketch > Format > Centerline** 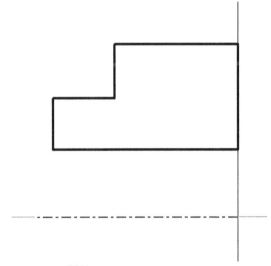.

6. Click **Line** / on the **Create** panel.

7. Create a centerline, as shown below.

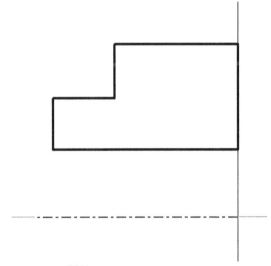

8. Click **Fix** 🔒 on the **Constrain** panel.

9. Select the Line 1.

10. Click **Dimension** on the **Constrain** panel.

11. Select the centerline and Line 2; a dimension appears.

12. Move the pointer horizontally toward the right and click to place the dimension.

13. Place the dimension and enter **4** in the **Edit Dimension** box.

14. Click the green check ✔ on the **Edit Dimension** dialog.

15. Select the centerline and Line 4; a dimension appears.

16. Move the pointer horizontally toward left and click to place the dimension.

17. Enter **2.4** in the **Edit Dimension** box.

18. Click the green check ✔ on the **Edit Dimension** dialog.

19. Select the centerline and Line 6; a dimension appears.

20. Move the pointer horizontally toward left and click to place the dimension.

21. Enter **1.2** in the **Edit Dimension** box.

22. Click the green check ✔ on the **Edit Dimension** dialog.

23. Create a dimension between Line 1 and Line 3.

24. Set the dimension value to 0.8 inches.

25. Create a dimension between Line 1 and Line 5.

26. Set the dimension value to 2 inches.

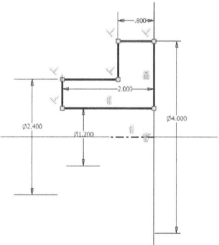

📝 *You can display all the constraints by right-clicking and selecting Show All Constraints option. You can hide all the constraints by right-clicking and selecting the Hide All Constraints option.*

27. Right-click and select **Finish 2D Sketch**.

Creating the Revolved Feature

1. On the ribbon, click **3D Model > Create > Revolve** (or) right-click and select **Revolve** from the Marking menu.

2. Click the **Full** button under the **Behavior** section on the **Revolve** Properties panel.

3. Click **OK** to create the revolved feature.

Creating the Cut feature

1. On the **Navigation** pane, click the **Orbit** icon.

2. Press and hold the left mouse button and drag the mouse; the model is rotated.
3. Rotate the model such that its back face is visible.
4. Right click and select **OK**.
5. On the **3D Model** tab of the ribbon, click the **Show Panels** icon located at the right corner, and then check the **Primitives** option from the drop-down.

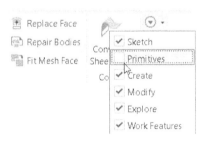

The **Primitives** panel is added to the ribbon.

6. On the ribbon, click **3D Model > Primitives > Primitive drop-down > Box** on the **Primitives** panel.

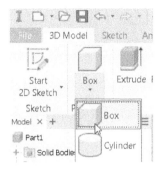

7. Click the back face of the part; the sketch starts.

8. Select the origin to define the center point.
9. Move the cursor diagonally toward the right.
10. Enter 4.1 in the horizontal dimension box.
11. Press Tab key and enter 0.472 in the vertical dimension box.

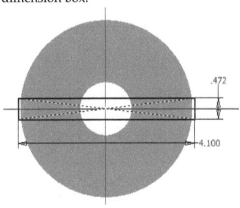

12. Press the Enter key; the **Extrude** Properties panel appears.
13. Expand the **Output** section on the **Extrude** Properties panel by clicking the **Output**.
14. Click the **Cut** 🔲 icon under the **Output** section on the **Extrude** Properties panel.
15. Enter 0.4 in the **Distance A** box.
16. Click the **Direction** > **Flipped** 🗹 icon under the **Behavior** section to reverse the direction.
17. Click **OK** to create the cut feature.

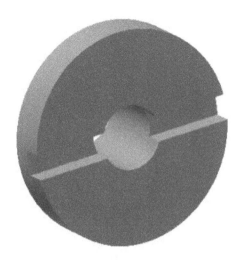

Creating another Cut feature

1. Click the **Home** icon located at the top left corner of the **ViewCube**.

2. Create a sketch on the front face of the base feature.

- On the ribbon, click **3D Model > Sketch > Start 2D Sketch**.
- Select the front face of the model.

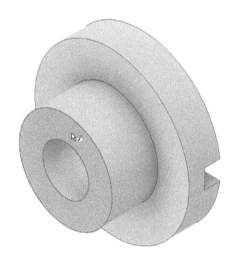

- Draw three lines and the circle, as shown in the figure.

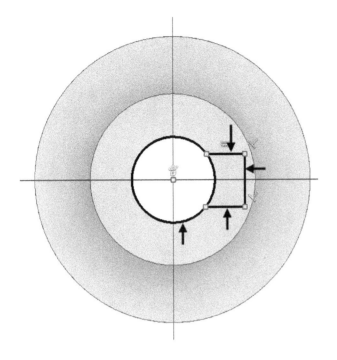

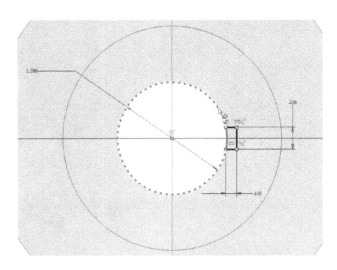

- Apply the **Horizontal** constraint to the horizontal lines, if not applied already.
- Apply the **Equal** constraint between the horizontal lines.
- Ensure that the endpoints of the horizontal line coincide with the circle.
- Apply dimension of 0.236 to the vertical line.
- Apply dimension of 0.118 to the horizontal line.
- Apply the dimension of the 1.2-inch diameter to the circle.

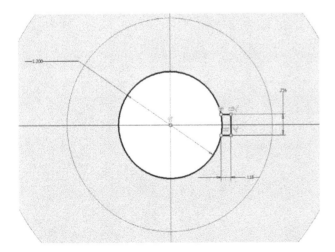

- On the ribbon, click **Sketch > Modify > Trim** ✂.
- Click on the circle to trim it.

3. Finish the sketch.

*You can hide or display the sketch dimensions. To do this, go to **View > Visibility > Object Visibility** and check the **Sketch Dimensions** option.*

4. Click **Extrude** on the **Create** panel of the **3D Model**.

5. Click on the region enclosed by the three lines and the arc.

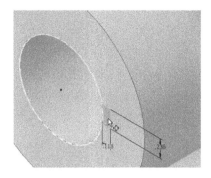

6. Click the **Through All** icon under the **Behavior** section.

7. Click the **Cut** [icon] icon under the **Output** section on the **Extrude** Properties panel.
8. Click **OK** to create the cut feature.

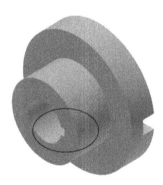

Adding a Fillet

1. On the ribbon, click **3D Model > Modify > Fillet** [icon] (or) right-click and select **Fillet** from the Marking menu.

2. Click on the inner circular edge.

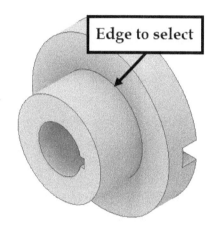

3. On the **Fillet** dialog, click the **Constant** tab, and then type 0.2 in the **Radius** box.
4. Click **OK** to add the fillet.

Saving the Part

1. Click **Save** [icon] on the **Quick Access Toolbar**.
2. On the **Save As** dialog, type-in **Flange** in the **File name** box.
3. Click **Save** to save the file.
4. Click **File Menu > Close**.

TUTORIAL 3

In this tutorial, you create the Shaft by performing the following:

- Creating a cylindrical feature
- Creating a cut feature

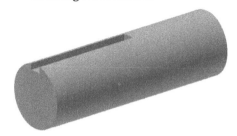

Starting a New Part File

1. On the ribbon, click **Get Started > Launch > New** 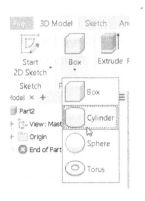.

2. On the **Create New File** dialog, select **Standard.ipt**.

3. Click **Create**.

Creating the Cylindrical Feature

1. On the ribbon, click **Primitives > Primitive drop-down > Cylinder**.

2. Click on the **XY** plane to select it; the sketch starts.

3. Click at the origin and move the cursor outward.

4. Enter 1.2 in the box attached to the circle.

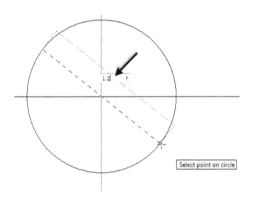

5. Press Enter key; the **Extrude** Properties panel appears.

6. Enter **4** in the **Distance A** box.

7. Click **OK** to create the cylinder.

Creating Cut feature

1. Create a sketch on the front face of the base feature.
   - On the ribbon, click **3D Model > Sketch > Start 2D Sketch**.
   - Select the front face of the cylinder.
   - On the ribbon, click **Sketch > Create > Line**.
   - Draw three lines, as shown.

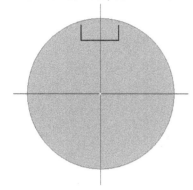

   - Apply the **Coincident** constraint between the endpoints of the vertical lines and the circular edge.
   - Add dimensions to the sketch.

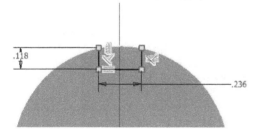

2. Finish the sketch.

3. Click **Extrude** on the **Create** panel.
4. Click on the region enclosed by the sketch.
5. Click the **Cut** icon on the **Extrude** Properties panel.
6. Set **Distance A** to **2.165**.
7. Click **OK** to create the cut feature.

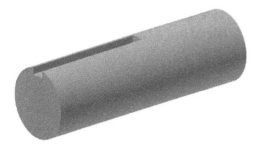

Saving the Part

1. Click **Save** on the **Quick Access Toolbar**; the **Save As** dialog appears.
2. Type-in **Shaft** in the **File name** box.
3. Click **Save** to save the file.
4. Click **File Menu > Close**.

TUTORIAL 4

In this tutorial, you create a Key by performing the following:

- Creating an Extruded feature
- Applying draft

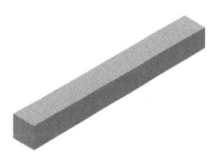

Start Extruded feature

1. Start a new part file using the **Standard.ipt** template.
2. On the ribbon, **Primitives > Primitive** drop-down **> Box**.
3. Select the XY plane.
4. Create the sketch, as shown in the figure.

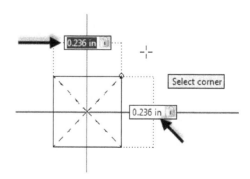

5. Press ENTER.
6. Enter 2 in the **Distance A** box.
7. Click **OK** to create the extrusion.

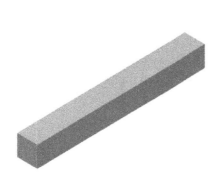

Applying Draft

1. On the ribbon, click **3D Model > Modify > Draft**.

2. Select the **Fixed Plane** option.

3. Select the front face as the fixed face.

4. Select the top face as the face to be draft.

5. Set **Draft Angle** to **1**.
6. Click the **Flip pull direction** button on the **Face Draft** dialog.

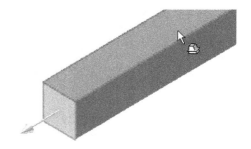

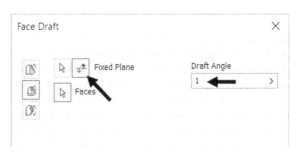

7. Click **OK** to create the draft.

Saving the Part

1. Click **Save** on the **Quick Access Toolbar**; the **Save As** dialog appears.
2. Type-in **Key** in the **File name** box.
3. Click **Save** to save the file.
4. Click **File Menu > Close**.

Chapter 3: Assembly Basics

In this chapter, you will:

- Add Components to assembly
- Apply constraints between components
- Check Degrees of Freedom
- Check Interference
- Create an exploded view of the assembly

TUTORIAL 1

This tutorial takes you through the creation of your first assembly. You create the Oldham coupling assembly:

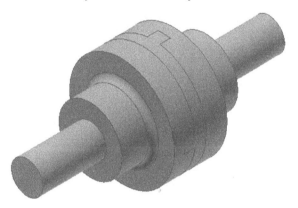

PARTS LIST		
ITEM	PART NUMBER	QTY
1	Disc	1
2	Flange	2
3	Shaft	2
4	Key	2

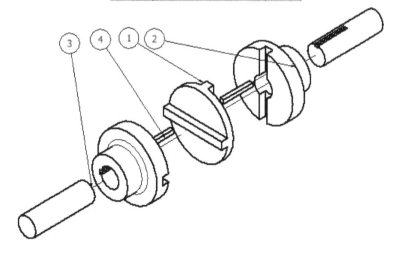

33

There are two ways of creating an assembly model.

- Top-Down Approach
- Bottom-Up Approach

Top-Down Approach

The assembly file is created first, and components are created in that file.

Bottom-Up Approach

The components are created first and then added to the assembly file. In this tutorial, you will create the assembly using this approach.

Starting a New Assembly File

1. To start a new assembly file, click the Assembly icon on the Home screen.

Inserting the Base Component

1. To insert the base component, click **Assemble > Component > Place from Content Center > Place** on the ribbon.

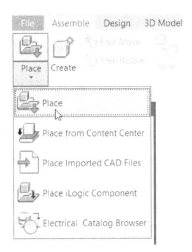

2. Browse to the project folder and double-click on **Flange.ipt**.

3. Right-click and select **Place Grounded at Origin**; the component is placed at the origin.

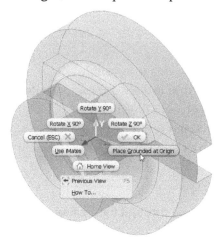

4. Right-click and select **OK**.

Adding the second component

1. To insert the second component, right-click and select **Place Component**; the **Place Component** dialog appears.

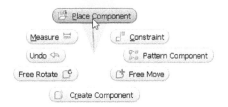

2. Browse to the project folder and double-click on **Shaft.ipt**.

3. Click in the window to place the component.

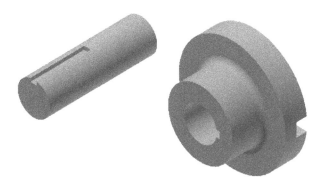

4. Right-click and select **OK**.

Applying Constraints

After adding the components to the assembly environment, you need to apply constraints between them. By applying constraints, you establish relationships between components.

1. To apply constraints, click **Assemble > Relationships > Constrain** on the ribbon.

The **Place Constraint** dialog appears on the screen.

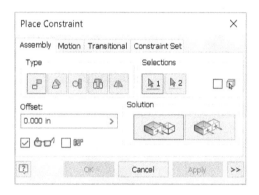

Different assembly constraints that can be applied using this dialog are given next.

Mate: Using this constraint, you can make two planar faces coplanar to each other.

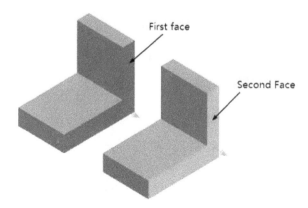

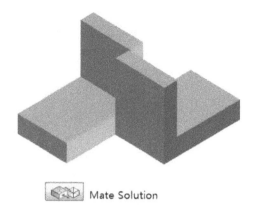

Mate Solution

Note that if you set the **Solution** to **Flush**, the faces will point in the same direction.

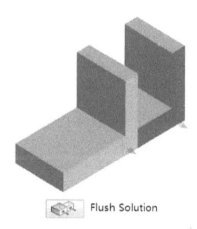

Flush Solution

You can also align the centerlines of the round faces. Select the two cylindrical faces to be aligned.

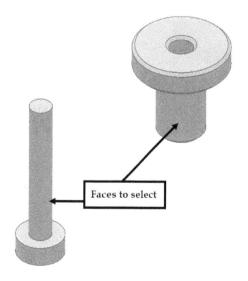

Select **Solution > Opposed** 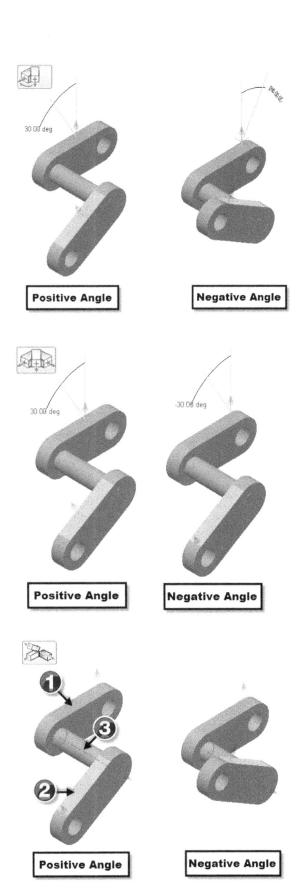 ; the axes of the selected round faces will be positioned in the direction opposite to each other.

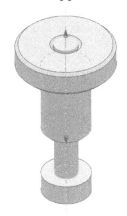

Select **Solution > Aligned** ; the axes of the selected round faces will be positioned in the same direction.

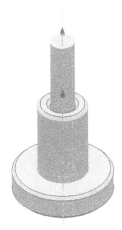

Select **Solution > Undirected** to the position component without specifying the axis direction.

Angle: Applies the angle constraint between two components.

Positive Angle **Negative Angle**

Positive Angle **Negative Angle**

Positive Angle **Negative Angle**

Tangent: This constraint is used to apply a tangent relation between two faces.

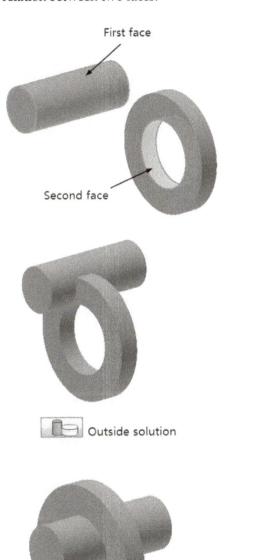

Outside solution

Inside Solution

Insert: This constraint is used to make two round faces coaxial. In addition, the planar faces of the cylindrical components will be on the same plane.

Check the **Lock Rotation** ⟲ option, if you want to lock the rotation of the component.

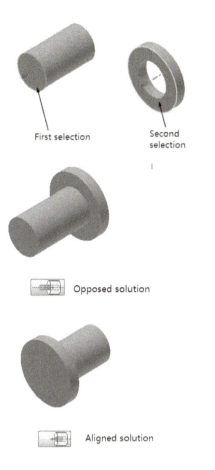

Symmetry: This constraint is used to position the two components symmetrically about a plane.

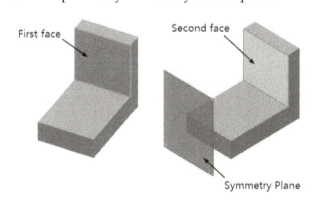

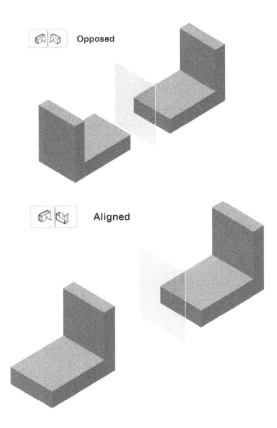

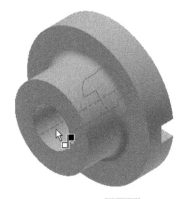

5. Click the **Aligned** icon on the **Place Constraint** dialog.

2. On the **Place Constraint** dialog, under the **Type** group, click the **Mate** icon.

3. Click on the round face of the Shaft.

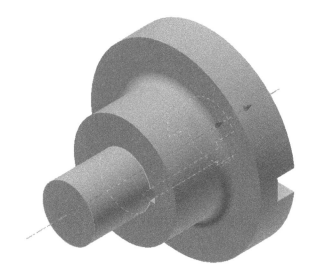

6. Click the **Apply** button.

7. Ensure that the **Mate** icon is selected in the **Type** group.

8. Click the **Free Orbit** icon on the Navigation Bar.

4. Click on the inner cylindrical face of the **Flange**.

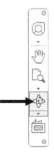

9. Move the pointer and place it on the model.

10. Click and drag the left mouse button toward left; the model is rotated such that the back side is displayed.
11. Right click and select **OK**.
12. Click the **Flush** button on the **Place Constraint** dialog.

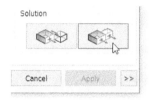

13. Click on the front face of the shaft.
14. Click on the slot face of the flange, as shown in the figure.

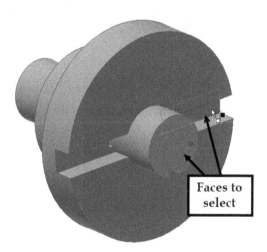

Faces to select

15. Click **Apply**. The front face of the Shaft and the slot face of the Flange are aligned.
16. Right click in the graphics window and select **Home View**.

17. Ensure that the **Mate** button is selected in the **Type** group.
18. Expand **Flange: 1** in the Browser window.
19. Select the XZ Plane of the Flange.

20. Expand **Shaft: 1** and select the YZ plane of the Shaft.

21. Click the **Flush** button on the **Place Constraint** dialog.

39

22. Click **OK** to assemble the components.

Adding the Third Component

1. To insert the third component, click **Assemble > Component > Place** on the ribbon.
2. Go to the project folder and double-click on **Key.ipt**.
3. Click in the graphics window to place the key.
4. Right-click and click **OK**.
5. Right-click on **Flange: 1** in the Browser window.
6. Click **Visibility** on the shortcut menu; the Flange is hidden.
7. Click **Constrain** on the **Relationships** panel.
8. Click **Mate** on the **Place Constraint** dialog.
9. Select **Mate** from the **Solution** group.

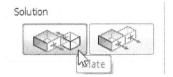

10. Click the right mouse button on the side face of the key and click **Select Other** on the shortcut menu.
11. Select the bottom face of the Key from the flyout.

12. Select the flat face of the slot.

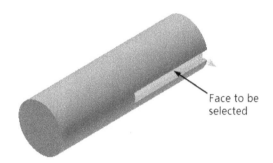

13. Click the **Apply** button. The bottom face of the key is aligned with the flat face of the slot.

14. Click the **Mate** icon on the **Place Constraint** dialog.
15. Select **Flush** from the **Solution** group.

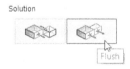

16. Select the front face of the Key and back face of the Shaft, as shown.

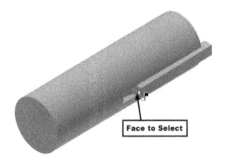

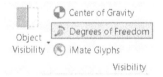

17. Click **Apply** on the dialog; the mate is applied.
18. Close the dialog.

   Now, you need to check whether the parts are fully constrained or not.

19. Click **View > Visibility > Degrees of Freedom** on the ribbon.

You will notice that an arrow appears pointing in the upward (or downward) direction. This means that the Key is not constrained in the Z-direction.

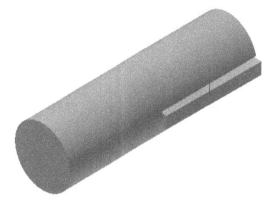

You must apply one more constraint to constrain the key.

20. Click **Constrain** on the **Relationships** panel of the **Assemble** ribbon tab.
21. Click the **Mate** icon the dialog.
22. Select **Flush** from the **Solution** group.
23. Expand the **Origin** node of the **Assembly** in the **Browser window** and select **XZ Plane**.

24. Expand the **Key: 1** node in the **Browser window** and select **YZ Plane**.

25. Click **OK**. The mate is applied between the two planes.

Now, you need to turn on the display of the Flange.

26. Right-click on the **Flange** in the **Browser window** and select **Visibility**; the **Flange** appears.

Checking the Interference

1. Click **Inspect > Interference > Analyze Interference** on the Ribbon. The **Interference Analysis** dialog appears.

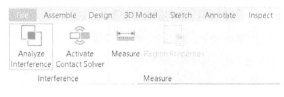

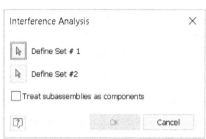

2. Select the Flange and Shaft as **Set #1**.
3. Click the **Define Set #2** button.
4. Select the Key as **Set # 2**.

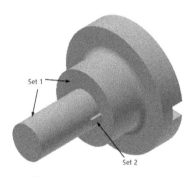

5. Click **OK**; the message box appears showing that there are no interferences.
6. Click **OK**.

Saving the Assembly

1. Click **Save** on the **Quick Access Toolbar**; the **Save As** dialog appears.
2. Type-in **Flange_subassembly** in the **File name box.**
3. Go to the project folder.
4. Click **Save** to save the file.
5. Click **File Menu > Close**.

Starting the Main assembly

1. On the ribbon, click **Get Started > Launch > New**.
2. On the **Create New File** dialog, click the **Standard.iam** icon.

Standard.iam

3. Click **Create** to start a new assembly.

Adding Disc to the Assembly

1. Click **Assemble > Component > Place** on the ribbon.
2. Go to the project folder and double-click on **Disc.ipt**.
3. Right-click and select **Place Grounded at Origin**; the component is placed at the origin.
4. Right-click and select **OK**.

Placing the Sub-assembly

1. To insert the sub-assembly, click the **Place** button on the **Component** panel of the ribbon.
2. Go to the project folder and double-click on **Flange_subassembly.iam**.
3. Click in the window to place the flange subassembly.
4. Right-click and click **OK**.

Adding Constraints

1. Click **Constrain** on the **Relationships** panel of the **Assemble** ribbon.
2. Click the **Insert** button on the **Place Constraint** dialog.

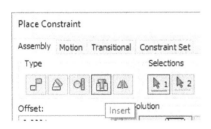

3. Select **Opposed** from the **Solution** group.

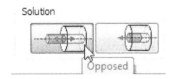

4. Click on the circular edge of the Flange.

5. Click on the circular edge of the Disc.

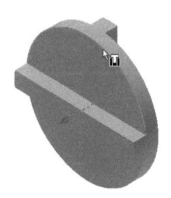

6. Click **OK** on the dialog.

Next, you have to move the subassembly away from the Disc to apply other constraints.

7. Click **Free Move** on the **Position** panel.

8. Select the flange subassembly and move it.

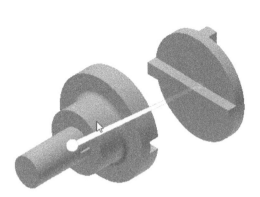

9. Click the **Constrain** button on the **Relationships** panel.
10. Click **Mate** on the **Place Constraints** dialog.
11. Select **Mate** from the **Solution** group.
12. Click the **View > Navigate > Orbit** on the ribbon.
13. Press and hold the left mouse button and drag the cursor toward left.
14. Release the mouse button, right click, and select **OK**.
15. Click on the face on the Flange, as shown in the figure.

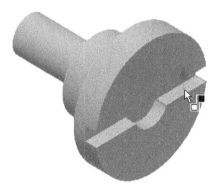

16. Click on the face on the Disc, as shown in the figure.

17. Click **OK** on the dialog.

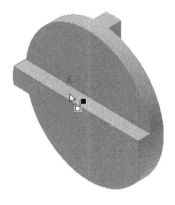

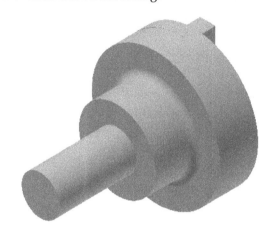

Placing the second instance of the Sub-assembly

1. Insert another instance of the Flange subassembly.
2. Apply the **Insert** and **Mate** constraints.

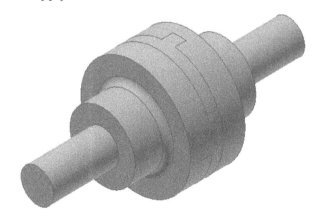

Saving the Assembly

1. Click **Save** on the **Quick Access Toolbar**; the **Save As** dialog appears.
2. Type-in **Oldham coupling** in the **File name** box.

3. Click **Save** to save the file.
4. Click **File Menu > Close**.

TUTORIAL 2

In this tutorial, you create the exploded view of the assembly:

Starting a New Presentation File

1. On the Home screen, click the Presentation icon (or) click **Get Started > Launch > New**, and then

select the Standard.ipn template from the **Create New File** dialog.

The Insert dialog appears.

2. On the **Insert** dialog, go to the project folder and double-click on the Oldham Coupling.iam file.

The Presentation Environment appears, as shown.

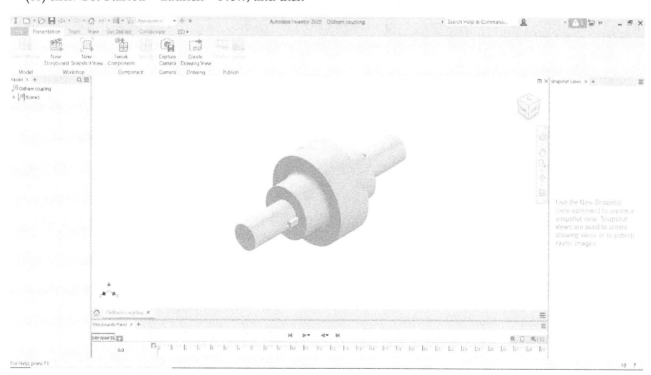

Creating a Storyboard Animation

1. In the **Model** tree, double-click on **Scene1** and type **Explosion**.

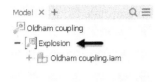

Before creating an exploded view, you need to take a look at the Storyboard displayed at the bottom of the window. The Storyboard has the Scratch Zone located at the left side of the timeline. Also, notice that the play marker is displayed at 0 seconds in the timeline.

2. Click in the Scratch Zone area and notice that the play marker is displayed inside it.

   Now, the changes made to the assembly in the Scratch Zone will be the starting point of the exploded animation. You can change the orientation of the assembly, hide a component, or change the opacity of the component. Use the **Capture Camera** tool on the **Camera** panel to set the camera position for the animation.

3. On the ribbon, click **View** tab > **Windows** panel > **User Interface** drop-down, and then check the **Mini toolbar** option. The Mini toolbar appears whenever you activate a tool.

4. Click the **Tweak Components** button on the **Component** panel of the **Presentation** ribbon tab. The mini toolbar appears with different options, as shown.

   Notice that the default duration for a tweak is 2.500 s. You can type a new value in the **Duration** box available on the Mini toolbar.

5. Select **Component** from the **Selection Filter** drop-down of the Mini toolbar.

6. Select the **All Components** from the Tracelines drop-down. This will create trace lines of all exploded components.

7. Select the Flange subassembly from the graphics window. The manipulator appears on the assembly.

   Now, you must specify the direction along which the sub-assembly will be exploded.

8. On the Mini toolbar, select **Local > World**.

9. Click the Z axis of the manipulator.

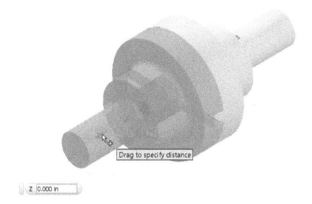

10. Type 4 in the Z box attached to the manipulator.

11. Click **OK** on the Mini toolbar.
12. Right click in the graphics window and select **Tweak Components** from the Marking menu.

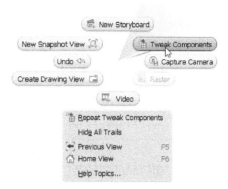

13. Select **Component** from the Selection filter drop-down on the Mini toolbar.
14. Select the other flange sub-assembly.
15. Click on the Z axis of the manipulator.
16. Type 4 in the Z box attached to the manipulator, and click **OK** .

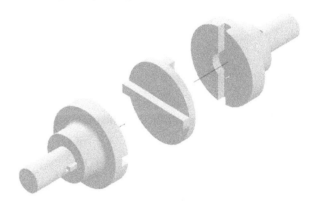

17. Click the **Tweak Components** button on the **Component** panel of the **Presentation** ribbon tab.
18. On the Mini Toolbar, select **Part** from the drop-down, as shown.

19. Select the front cylinder.

20. On the Mini toolbar, select **Local > World**.

21. Click on the Z axis of the manipulator.

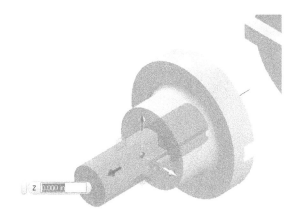

22. Type 4 in the box attached to the manipulator, and then press Enter.
23. Click **OK** on the Mini Toolbar.
24. Activate the **Tweak Components** command.
25. Zoom into the flange and click on the key, as shown.

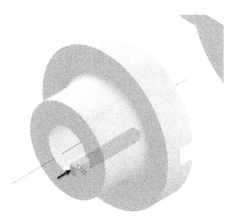

26. Select **Local > World** on the mini toolbar.
27. Click on the Z axis of the manipulator.

28. Type 3.15 in the box attached to the manipulator and press Enter.
18. Likewise, explode the parts of the flange subassembly in the opposite direction. The explosion distances are the same.

Animating the Explosion

1. To play the animation of the explosion, click the **Play Current Storyboard** button on the Storyboard.

2. Click the **Reverse Play Current Storyboard** button on the Storyboard.

You can publish the animation video using the **Video** tool available on the **Publish** panel.

3. Make sure that the play marker is at 0 secs on the timeline.

4. On the ribbon, click **Presentation** tab > **Publish** panel > **Video** .

5. On the **Publish to Video** dialog, select **Current Storyboard** option from the **Publish Scope** section.

    You can also select **Current Storyboard Range** and specify the start and end position of the storyboard.

6. On the **Publish to Video** dialog, click the folder icon and specify the project folder as the **File Location**.
7. Set the **File Format** to **WMV File (*.wmv)**.
8. Check the **Reverse** option to reverse the animation.
9. Leave the other default settings and click **OK**; **Publish Video Progress** dialog appears.

    A message box appears that the video has been published.

10. Click **OK** on the message box.

Taking the Snapshot of the Explosion

1. Click and drag the play marker on the timeline to 15 seconds.

    You can capture the snapshot of the current position of the assembly using the **New Snapshot View** tool.
2. On the ribbon, click **Presentation** tab >

    Workshop panel > **New Snapshot View** .

The snapshot appears in the Snapshot Views window. Notice that playmarker on the snapshot. It indicates that the snapshot is dependent on the storyboard.

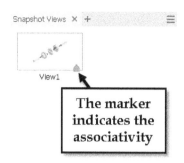

The marker indicates the associativity

For example, if you make changes to the assembly at the position of the playmarker where the snapshot was taken, the Update View symbol appears on the snapshot view. You need to click on the Update View symbol to update the snapshot.

3. Click **Save** on the **Quick Access Toolbar**; the **Save As** dialog appears.
4. Type-in **Oldham_coupling** in the **File name** box.
5. Go to the project folder.
6. Click **Save** to save the file.
7. Click **OK**.
8. Click **File Menu > Close**.

Chapter 4: Creating Drawings

In this chapter, you will generate 2D drawings of the parts and assemblies.

In this chapter, you will:

- Insert standard views of a part model
- Create centerlines and centermarks
- Retrieve model dimensions
- Add additional dimensions and annotations
- Create Custom Sheet Formats and Templates
- Insert exploded view of the assembly
- Insert a bill of materials of the assembly
- Apply balloons to the assembly

TUTORIAL 1

In this tutorial, you will create the drawing of Flange.ipt file created in the second chapter.

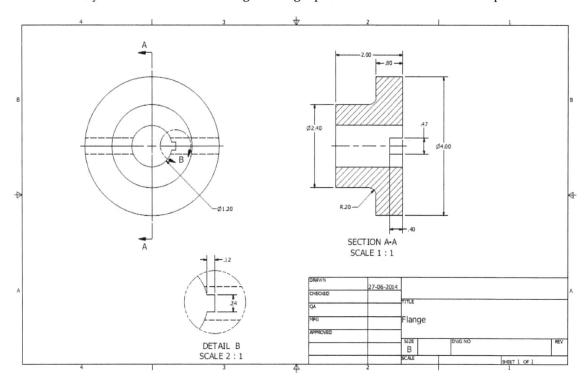

Starting a New Drawing File

1. To start a new drawing, click the **Drawing** icon on the Home screen.

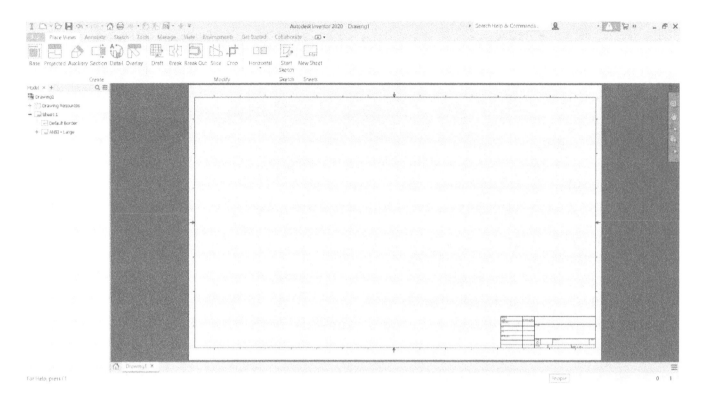

Editing the Drawing Sheet

1. To edit the drawing sheet, right-click on **Sheet:1** in the **Browser window** and select **Edit Sheet** from the shortcut menu.

2. On the **Edit Sheet** dialog, set **Size** to **B**.

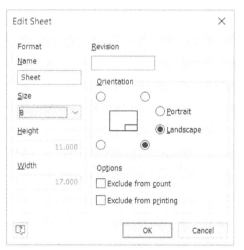

3. Click **OK**.

The drawing views in this tutorial are created in the Third Angle Projection. If you want to change the type of projection, then following the steps given next:

4. Click **Manage > Styles and Standards > Style Editor** on the ribbon.

Creating Drawings

5. On the **Style and Standard Editor** dialog, specify the settings shown in the figure.

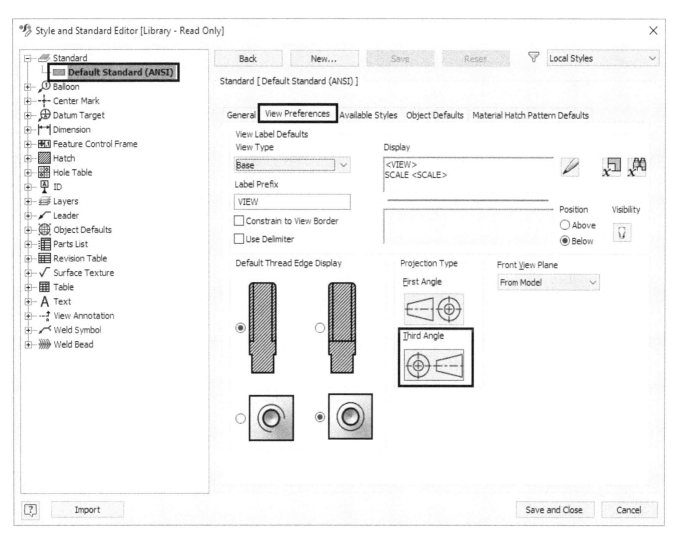

6. Click **Save and Close**.

Generating the Base View

1. To generate the base view, click **Place Views > Create > Base** on the ribbon.

2. On the **Drawing View** dialog, click **Open existing file** .
3. On the **Open** dialog, browse to the project folder.

4. Set **Files of type** to **Inventor Files (*.ipt, *.iam, *.ipn)**, and then double-click on **Flange.ipt**.
5. Set the **Style** to **Hidden Line** .
6. Set **Scale** to **1:1**.
7. Click on the preview, drag, and place it at the left side on the drawing sheet, as shown.
8. Click **OK** on the dialog.

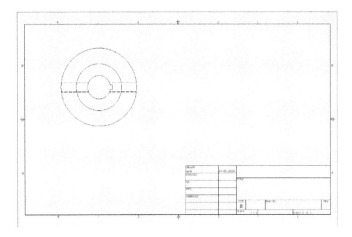

Generating the Section View

1. To create the section view, click **Place Views > Create > Section** on the ribbon.

2. Select the base view.
3. Place the cursor on the top quadrant point of the circular edge, as shown.

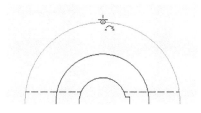

4. Move the pointer upward and notice the dotted line.

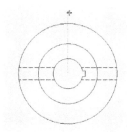

5. Click on the dotted line and move the cursor vertically downwards.
6. Click outside the bottom portion of the view, as shown.

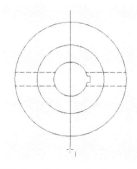

7. Right-click and select **Continue**.

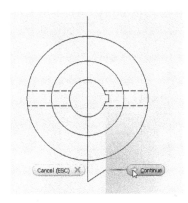

8. Move the cursor toward the right and click to place the section view.

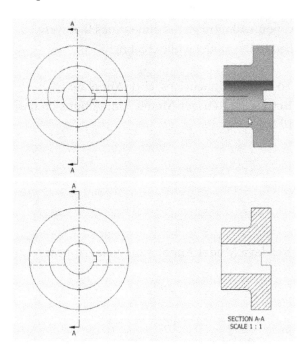

Creating the Detailed View

Now, you have to create a detailed view of the keyway, which is displayed, in the front view.

1. To create a detailed view, click **Place Views > Create > Detail** on the ribbon.

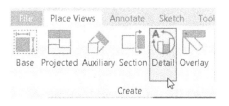

2. Select the base view.
3. On the **Detail View** dialog, specify the settings, as shown next.

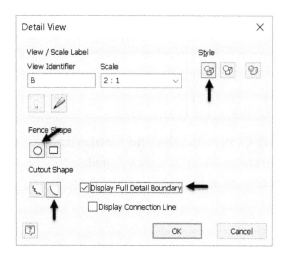

4. Specify the center point and boundary point of the detail view, as shown in the figure.

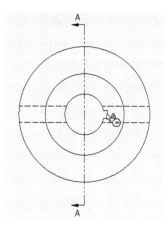

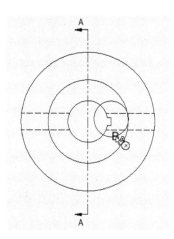

5. Place the detail view below the base view.

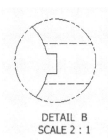

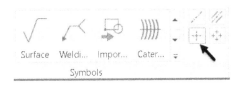

DETAIL B
SCALE 2 : 1

Creating Center marks and Centerlines

1. To create a center mark, click **Annotate > Symbols > Center Mark** on the ribbon.

2. Click on the outer circle of the base view.

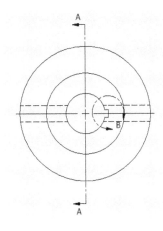

3. To create a centerline, click **Annotate > Symbols > Centerline Bisector** on the ribbon.

4. Click on the inner horizontal edges of the section view.

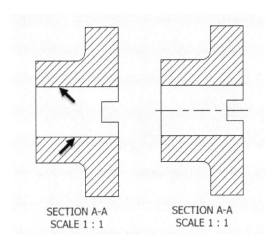

SECTION A-A   SECTION A-A
SCALE 1 : 1   SCALE 1 : 1

Retrieving Dimensions

Now, you will retrieve the dimensions that were applied to the model while creating it.

1. To retrieve dimensions, click **Annotate > Retrieve > Retrieve Model Annotations** on the ribbon.

The **Retrieve Model Annotation** dialog appears.

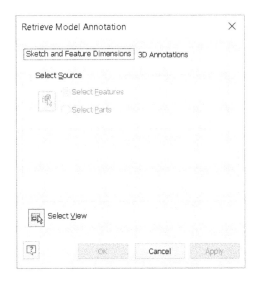

2. Select the section view from the drawing sheet.

Now, you must select the dimensions to retrieve.

3. Drag a window on the section view to select all the dimensions.

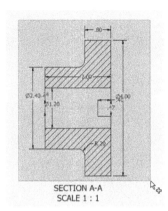

4. Click **Select Features** under the **Select Source** group.
5. Click **OK** to retrieve feature dimensions.

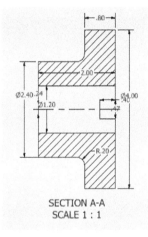

6. Click **Annotate > Dimension > Arrange** on the ribbon.

7. Drag a selection box and select all the dimensions of the section view.

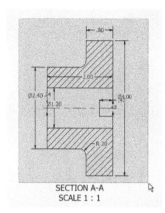

8. Click the right mouse button and select **OK**.

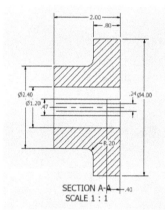

9. Select the unwanted dimensions and press Delete.

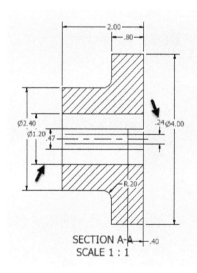

10. Click and drag the dimensions to arrange them correctly.

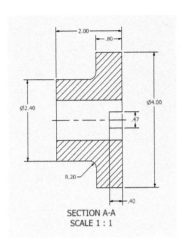

SECTION A-A
SCALE 1 : 1

Adding additional dimensions

1. To add dimensions, click **Annotate > Dimension > Dimension** on the ribbon.

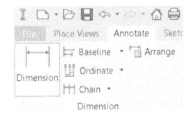

2. Select the center hole on the base view.

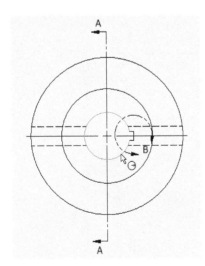

3. Right-click and select **Dimension Type > Diameter**.

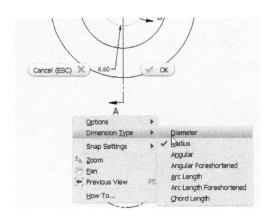

4. Place the dimension, as shown in the figure. The **Edit Dimension** dialog appears.

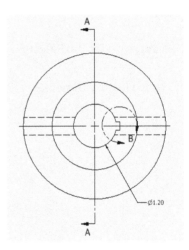

5. Click **OK**.
6. Create the dimensions on the detail view, as shown in the figure.

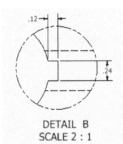

DETAIL B
SCALE 2 : 1

Populating the Title Block

1. To populate the title block, click **File Menu > iProperties**.

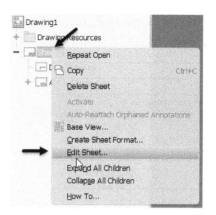

2. On the **Flange iProperties** dialog, click the tabs one-by-one and type-in data in respective fields.
3. Click **Apply** and **Close**.

Saving the Drawing

1. Click **Save** on the **Quick Access Toolbar**; the **Save As** dialog appears.
2. Type-in **Flange** in the **File Name** box.
3. Go to the project folder.
4. Click **Save** to save the file.
5. Click **File Menu > Close**.

TUTORIAL 2

In this tutorial, you will create a custom template, and then use it to create a new drawing.

Creating New Sheet Format

1. On the ribbon, click **Get Started > Launch > New**.
2. On the **Create New File** dialog, click the **Standard.idw** icon.

Standard.idw

3. Click **Create** to start a new drawing file.
4. To edit the drawing sheet, right-click on **Sheet:1** in the **Browser window** and select **Edit Sheet** from the shortcut menu.

5. On the **Edit Sheet** dialog, set **Size** to **B**.

Under the **Orientation** section, you can change the orientation of the title block as well as the sheet orientation.

6. Click **OK**.
7. In the **Browser window**, expand the **Drawing Resources > Sheet Formats** folder to view different sheet formats available. Now, you will add a new sheet format to this folder.
8. Click the right mouse button on the **Borders** folder and select **Define New Border**.

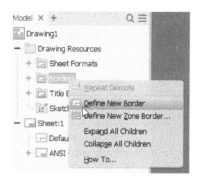

Now, you can create a new border using the sketch tools available in the **Sketch** tab.

9. Click **Finish Sketch** on the **Sketch** tab of the ribbon.
10. On the **Border** dialog, click **Discard**.
11. In the Browser window, click the right mouse button on the **Borders** folder and select **Define New Zone Border**.

12. On the **Default Drawing Border Parameters** dialog, type-in **4** in the **Vertical Zones** box and click **OK**.

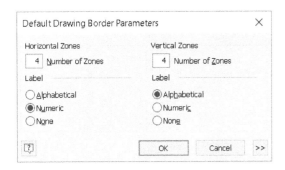

13. Click **Finish Sketch** on the **Sketch** tab of the ribbon.
14. On the **Border** dialog, type-in **4-Zone Border** and click **Save**.

15. Expand the **Title Blocks** folder and click the right mouse button on **ANSI-Large**.
16. Select **Edit** from the shortcut menu.

17. On the **Sketch** tab of the ribbon, click **Insert > Image**.

18. Draw a rectangle in the **Company** cell of the title block. This defines the image size and location.

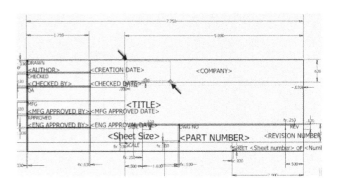

19. Go to the location of your company logo or any other image location. You must ensure that the image is located inside the project folder.
20. Select the image file and click **Open**. This will insert the image into the title block.

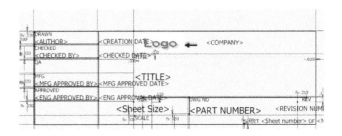

21. Click **Finish Sketch** on the ribbon.
22. Click **Save As** on the **Save Edits** dialog.
23. Type-in **ANSI-Logo** in the **Title Block** dialog.

24. Click **Save**.
25. In the Browser window, expand **Sheet:1** and click the right mouse button on **Default Border**.
26. Select **Delete** from the shortcut menu.

27. Expand the **Borders** folder and click the right mouse button on **4-Zone Border**.
28. Select **Insert** from the shortcut menu.

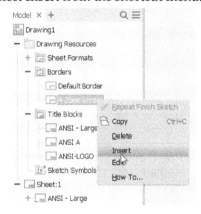

29. Click **OK** on the **Edit Drawing Border Parameters** dialog.
30. Expand **Sheet:1** and click the right mouse button on **ANSI-Large**.
31. Select **Delete** from the shortcut menu.

32. Expand the **Title Blocks** folder and click the right mouse button on **ANSI-Logo**.
33. Select **Insert** from the shortcut menu to insert the title block.

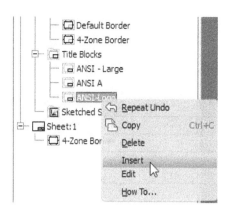

34. Click the right mouse button on **Sheet:1** and select **Create Sheet Format**.

35. Type-in **Custom Format** in the **Create Sheet Format** dialog, and then click **OK**.

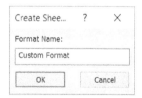

You will notice that the new sheet format is listed in the **Sheet Formats** folder.

61

Creating a Custom Template

1. On the ribbon, click **Tools > Options > Document Settings**.

On **Document Settings** dialog, you can define the standard, sheet color, drawing view settings, and sketch settings.

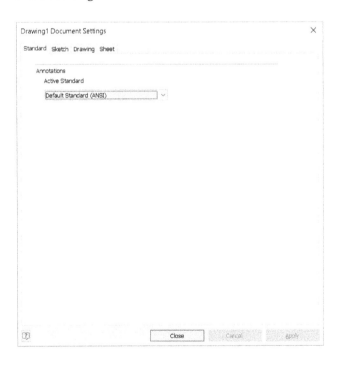

2. Leave the default settings on this dialog and click **Close**.
3. In the Browser window, expand the **Sheet Formats** folder and double-click on **Custom Format**.
4. Click the right mouse button on **Sheet: 2** and select **Delete Sheet** from the shortcut menu.

5. Click **OK**.
6. On the ribbon, click **Manage > Styles and Standards > Styles Editor**.
7. On the **Style and Standard Editor** dialog, select **Dimension > Default (ANSI)**.

8. Click the **New** button located at the top of the dialog.
9. Type-in **Custom Standard** in the **New Local Style** dialog, and then click **OK**.
10. Click the **Units** tab and set **Precision** to **3.123**.

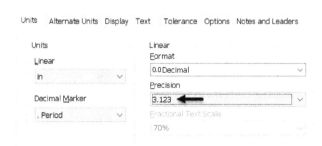

11. Click **Save and Close**.
12. On the **File Menu**, click **Save As > Save Copy As Template**. This will take you to the templates folder on your drive.
13. Type-in **Custom Template** in the **File name** box.
14. Click **Save**.
15. Close the drawing file without saving it.

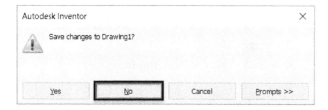

Starting a Drawing using the Custom template

1. On the ribbon, click **Get Started > Launch > New**.
2. On the **Create New File** dialog, click the **Custom Standard.idw** icon.

Custom
Template.idw

3. Click **Create** and **OK** to start a new drawing file.

Generating the Drawing Views

1. To generate views, click **Place Views > Create > Base** on the ribbon.
2. On the **Drawing View** dialog, click **Open existing file**.
3. Go to the project folder and double-click on **Disc.ipt**.
4. Select **Front** from the ViewCube displayed on the sheet.

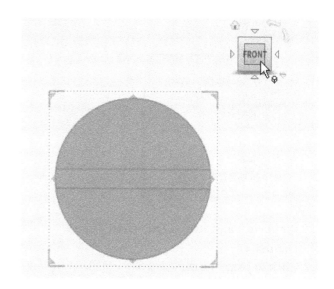

5. Set **Scale** to **1:1**.
6. Click and drag the view to top-center of the drawing sheet.
7. Move the cursor downwards and click to place the projected view.
8. Right-click and select **OK**.

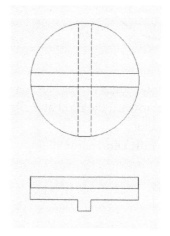

Adding Dimensions

1. On the ribbon, click **Annotate > Dimension > Dimension**.
2. Select the circular edge on the base view.
3. On the ribbon, click **Annotate > Format > Select Style > Custom Standard**.

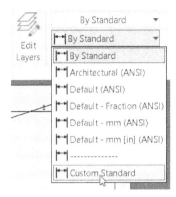

4. Right-click and select **Dimension Type >
   Diameter**.
5. Click to place the dimension.
6. Click **OK**.

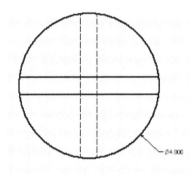

7. Select the horizontal edges on the base view.
8. Move the pointer toward the right and click to
   place the dimension.
9. Click **OK** on the **Edit Dimension** dialog.

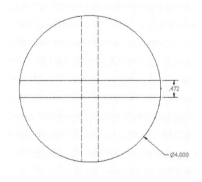

10. Add other dimensions to drawing.

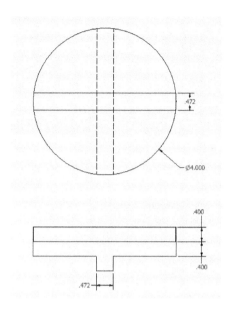

11. Right-click and select **OK** to deactivate the
    Dimension tool.
12. Save and close the drawing file.

TUTORIAL 3

In this tutorial, you will create the drawing of the
Oldham coupling assembly created in the previous
chapter.

Creating a New Drawing File

1. On the ribbon, click **Get Started > Launch >
   New**.
2. On the **Create New File** dialog, click the **Custom
   Standard.idw** icon.

Custom
Template.idw

3. Click **Create** and **OK** to start a new drawing file.

Generating Base View

1. To generate the base view, click **Place Views >
   Create > Base** on the ribbon; the **Drawing
   View** dialog appears.
2. Click **Open existing file** on this dialog; the
   Open dialog appears.
3. Go to the project folder and double-click on
   Oldham_Coupling.iam.

4. Click the **Home** icon located above the ViewCube.

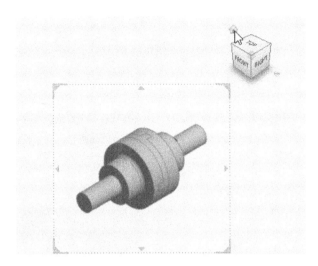

5. Set **Scale** to **1/2**.
6. Click and drag the view to the top left corner.
7. Right-click and select **OK**.

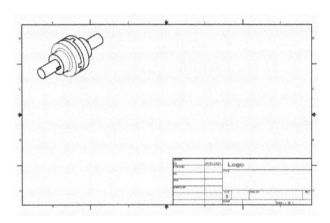

Generating the Exploded View

1. To generate the base view, click **Place Views > Create > Base** on the ribbon; the **Drawing View** dialog appears.
2. Click **Open existing file** on this dialog; the **Open** dialog appears.
3. Go to the project folder and double-click on **Oldham_Coupling.ipn**.
4. Click the **Home** icon located above the ViewCube.
5. Set **Scale** to **1/2**.
6. Click and drag the view to the center of the drawing sheet.

7. Right-click and select **OK**.

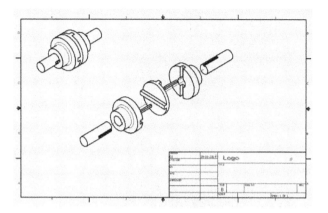

Configuring the Parts list settings

1. Click **Manage > Styles and Standards > Style Editor** on the ribbon; the **Style and Standard Editor** dialog appears.
2. Expand the **Parts List** node and select **Parts List (ANSI)**.

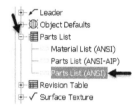

3. Click the **Column Chooser** button under the **Default Columns Settings** group; the **Parts List Column Chooser** dialog appears.

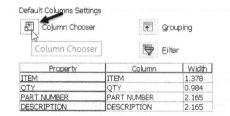

4. On this dialog, select **DESCRIPTION** from the **Selected Properties** list and click the **Remove** button.
5. Select **PART NUMBER** from the **Selected Properties** list and click **Move Up**.
6. Click **OK**.
7. Click **Save and Close**.

Creating the Parts list

1. To create a parts list, click **Annotate > Table > Parts List** on the ribbon; the **Parts List** dialog appears.

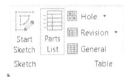

2. Select the exploded view from the drawing sheet.
3. Select **Parts Only** from the **BOM View** drop-down under the **BOM Settings and Properties** group.

4. Click **OK** twice.
5. Place the part list above the title block.

PARTS LIST		
ITEM	PART NUMBER	QTY
1	Disc	1
2	Flange	2
3	Shaft	2
4	Key	2

Creating Balloons

1. To create balloons, click **Annotate > Table > Balloon > Auto Balloon** on the ribbon; the **Auto-Balloon** dialog appears.

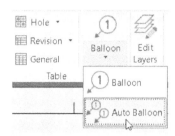

2. Select the exploded view from the drawing sheet.
3. Select all the parts in the exploded view.
4. Select **Horizontal** from the **Placement** group.
5. Click the **Select Placement** button in the **Placement** group.

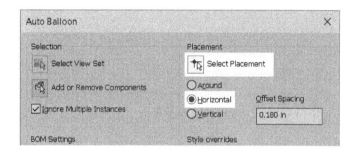

6. Click above the exploded view.
7. Click **OK** to place the balloons.

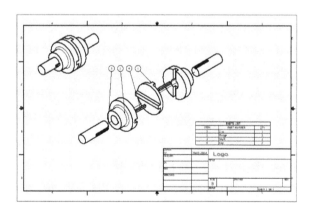

Saving the Drawing

1. Click **Save** on the **Quick Access Toolbar**; the **Save As** dialog appears.
2. Type-in **Oldham_Coupling** in the **File Name** box.
3. Go to the project folder.
4. Click **Save** to save the file.
5. Click **OK**.
6. Click **File Menu > Close**.

Chapter 5: Sketching

In this chapter, you will learn the sketching tools. You will learn to create:

- Rectangles
- Polygons
- Resolve Sketch
- Relations
- Splines
- Ellipses
- Move, rotate, scale, copy, and stretch entities
- Circles
- Arcs
- Circular pattern
- Trim Entities
- Fillets and Chamfers

Creating Rectangles

A rectangle is a four-sided object. You can create a rectangle by just specifying its two diagonal corners. However, there are various tools to create a rectangle. You can access these tools from the **Rectangle** drop-down available on the **Create** panel of the **Sketch** ribbon. These tools are explained next.

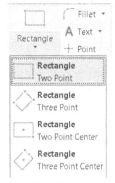

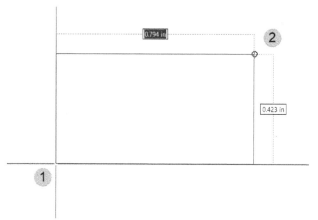

1. On the **My Home** page, click **New > Part**.
2. On the ribbon, click **3D Model > Sketch > 2D Sketch** and select the XY plane.
3. On the ribbon, click **Sketch > Create > Rectangle Drop-down > Rectangle Two Point**.
4. Select the origin point to define the first corner.
5. Move the pointer diagonally and click to define the second corner.

6. On the ribbon, click the **Sketch > Create > Rectangle Two Point Center**.
7. Click in the graphics window to define the center point.
8. Move the pointer and click to define the corner.

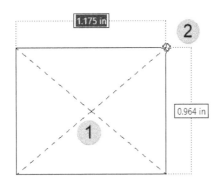

9. On the ribbon, click the **Sketch > Create > Rectangle Drop-down > Rectangle Three Point** . This option creates a slanted rectangle.

10. Select two points to define the width and inclination angle of the rectangle.

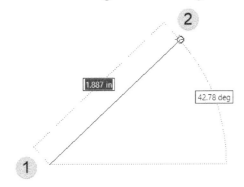

11. Select the third point to define its height.

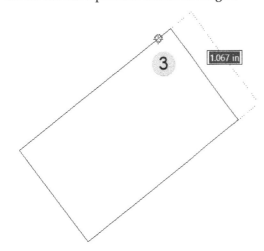

12. On the ribbon, click **Sketch > Create > Rectangle Drop-down > Rectangle Three Point Center Rectangle** .

13. Click to define the center point of the rectangle.

14. Move the pointer and click to define the midpoint of one side.

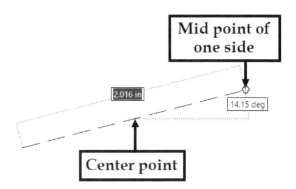

15. Move the pointer and click to define the height.

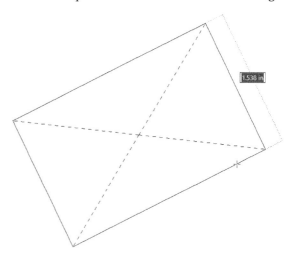

Creating Polygons

A Polygon is a shape having many sides ranging from 3 to 1024. In Inventor, you can create regular polygons having sided with equal length. Follow the steps given next to create a polygon.

1. Start a 2D sketch.

2. On the ribbon, click **Sketch > Create > Polygon** .

3. On the **Polygon** dialog, type **8** in the **Number of Sides** box.

4. Select **Inscribed circle** . This option creates a polygon with its vertices touching an imaginary circle. You can also select the **Circumscribed**

68

circle 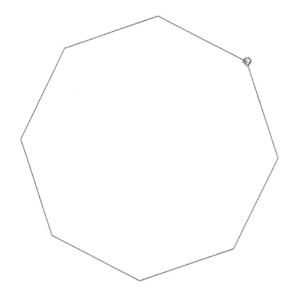 option to create a polygon with its sides touching an imaginary circle.

5. Click to define the center of the polygon.
6. Move the pointer and click to define the size and angle of the polygon.

7. Click **Done** to deactivate the tool.

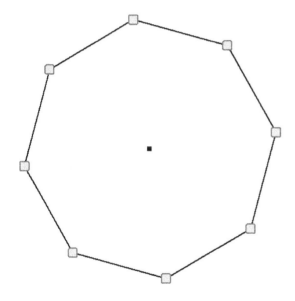

Driving and Driven Dimensions

When creating sketches for a part, Inventor will not allow you to over-constrain the geometry. The term 'over-constrain' means adding more dimensions than required. The following figure shows a fully constrained sketch. If you add another dimension to this sketch (e.g., diagonal dimension), the **Autodesk Inventor** dialog appears.

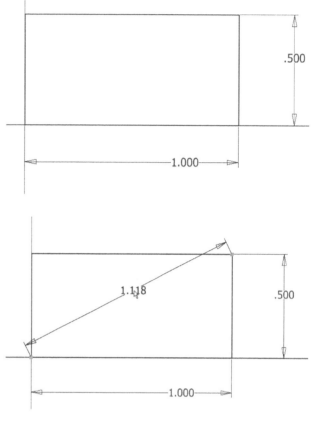

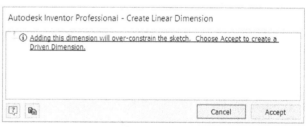

Now, click **Accept** to convert the new dimension into a driven dimension.

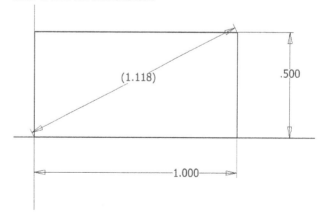

Now, if you change the value of the width, the reference dimension along the diagonal updates, automatically.

Constraints

Constraints are used to control the shape of a sketch by establishing relationships between the sketch elements. You can add constraints using the tools available on the **Constrain** panel.

Coincident Constraint

This constraint connects a point to another point.

1. On the **Sketch** tab of the Ribbon, click **Constrain > Coincident Constrain** .
2. Select two points. The selected points are connected together.

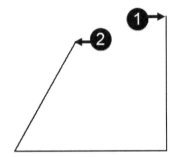 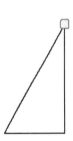

Horizontal Constraint

To apply the **Horizontal** Constraint, click on a line and click the **Horizontal Constraint** icon on the context toolbar.

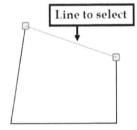 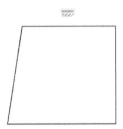

You can also align two points or vertices horizontally. Click the **Horizontal Constraint** icon on the Constrain panel, and then select the two points.

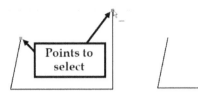

Vertical Constraint

Use the **Vertical Constraint** to make a line vertical. You can also use this constraint to align two points or vertices vertically.

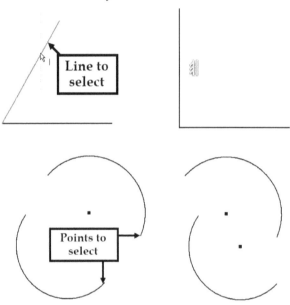

Tangent Constraint

This constraint makes an arc, circle, or line tangent to another arc or circle. Click the **Tangent Constraint** icon on the **Constrain** panel. Select a circle, arc, or line. Next, select another circle or arc; the two elements will be tangent to each other.

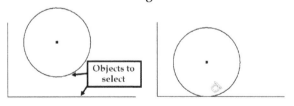

Parallel Constraint

Use the **Parallel Constraint** to make two lines parallel to each other. To do this, click the **Parallel**

Constraint icon on the **Constrain** panel. Next, select the two lines to be parallel.

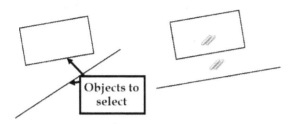

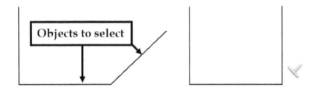

Perpendicular Constraint

Use the **Perpendicular Constraint** to make two entities perpendicular to each other.

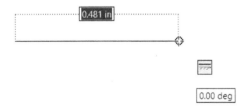

Automatic Constraints

Inventor automatically adds constraints when you create sketch elements.

1. Start a new sketch and activate the **Line** tool from the **Sketch** ribbon.
2. Click to specify the **start** point of the line.
3. Move the pointer in the horizontal direction and notice the **Horizontal** constraint flag.
4. Click to create a line with the **Horizontal** constraint.

5. Move the pointer vertically in the upward direction and notice the **Perpendicular** constraint flag.
6. Click to create a line with the **Perpendicular** constraint.

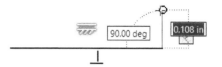

7. Create an inclined line, as shown.

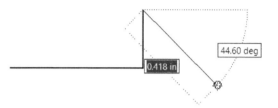

8. Move the pointer along the inclined line and notice the **Collinear** constraint flag.

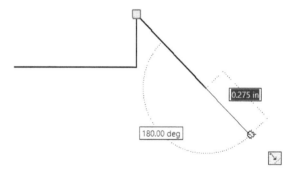

9. On the ribbon, click **Sketch** tab > **Create** panel > **Circle**, and then create a circle.
10. Activate the **Line** tool and click on the circle.
11. Move the pointer around the circle and notice that the line maintains the **Tangent** constraint with the circle.

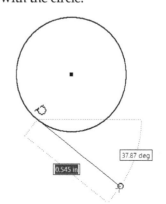

12. Click to create a line which is tangent and coincident to the circle. Right click and select **OK**.

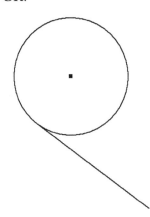

Deleting Constraints

You can delete constraints by using the following methods.

- On the status bar, click the **Show Constraints** icon.
- Select the constraint and press **Delete** on your keyboard.

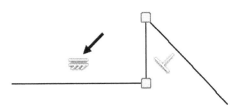

Hiding Constraints

To hide sketch constraints, click the **Hide All Constraints** icon located on the status bar.

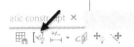

Project Geometry

The **Project Geometry** tool helps you to use the edges of the part geometry to create sketch elements.

1. Download the Chapter 5: Sketching folder from the Companion website.
2. Open the Project_geomtry part file from the folder.

3. Start a sketch on the plane offset to the model face, as shown.
4. On the **Sketch** ribbon, click **Modify > Project Geometry** .
5. Place the pointer on any one of the edges.
6. Click the down-arrow next to fly out, and then select **Loop**.

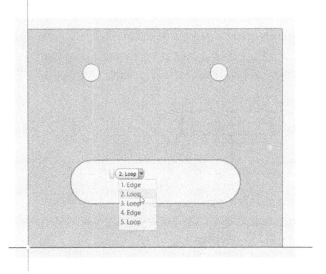

The loop is projected onto the sketch plane. The projected edges are displayed in yellow color.

7. Click on the round edges of the model.

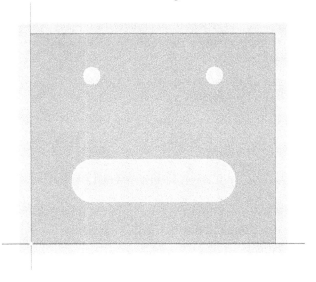

Tutorial 1

In this tutorial, you will create a sketch using the **Line**, **Three Point Arc**, and **Circle** tools.

1. Activate the **Sketch** Environment.
2. On the ribbon, click **Sketch > Create >Line.**
3. Click on the origin point to define the start point of the line.
4. Move the pointer horizontally toward the right and click.

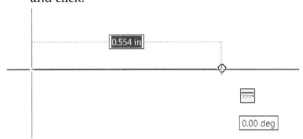

5. Move the pointer upward and click when a dotted line appears.

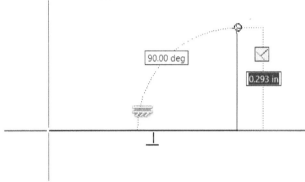

6. Move the pointer away from the endpoint of the vertical line.
7. Again, place the pointer on the end point of the vertical line.
8. Press and hold the left mouse button.
9. Move the pointer diagonally toward the top left corner and click. Next, right click and select **OK**.

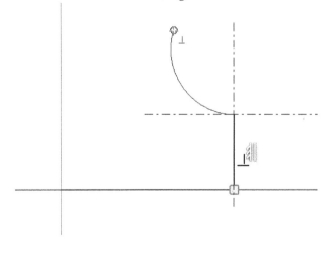

10. On the ribbon, click **Sketch > Format > Construction** .
11. Activate the **Line** tool.
12. Select the origin point of the sketch, move the pointer vertically upwards, and click. Press Esc.

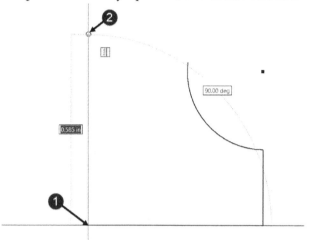

13. On the ribbon, click **Sketch > Pattern > Mirror** .
14. Select the arc, horizontal, and vertical lines.
15. On the **Mirror** dialog, click the **Mirror line** button and select the vertical construction line. Click **Apply** and **Done**.

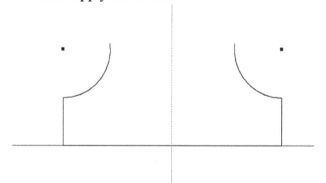

16. On the ribbon, click **Sketch > Format > Construction** .
17. On the ribbon, click **Sketch** tab > **Create** panel > **Arc Three Point** .
18. Select the end point of the left and right arc, and then move the pointer upward.
19. Click when the **Tangent** constraint appears.

73

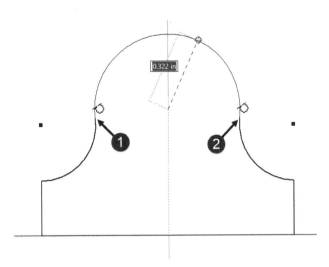

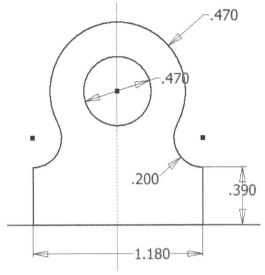

20. Right click and select **OK**.
21. On the ribbon, click **Sketch** tab > **Create** panel > **Circle Center Point**.
22. Select the centerpoint of the three-point arc, move the pointer outward, and click to create the circle.

24. Click **Finish Sketch**.

Ellipses

Ellipses are also non-uniform curves, but they have a regular shape. They are actually splines created in regular closed shapes.

1. Activate the **Sketch** environment.
2. On the ribbon, click **Sketch** tab > **Create** panel > **Circle** drop-down > **Ellipse** .

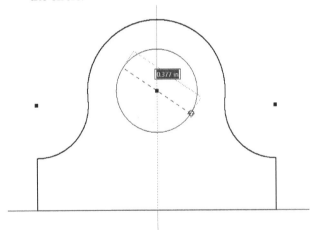

3. Pick a point in the graphics window to define the center of the ellipse.
4. Move the pointer and click to define the radius and orientation of the first axis.

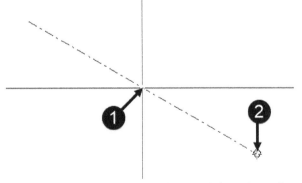

23. Add dimensions to the sketch, as shown.

5. Move the pointer and click to define the radius of the second axis.

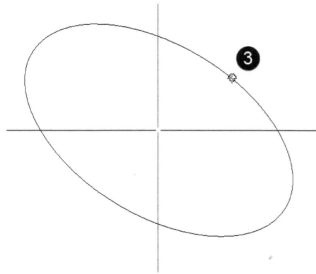

Click and drag the ellipse the notice that it is under-defined. You need to add dimensions and constraints to define the ellipse fully.

6. Activate the **Line** tool and select the centerpoint of the ellipse.
7. Move the pointer upward and select a point on the ellipse.

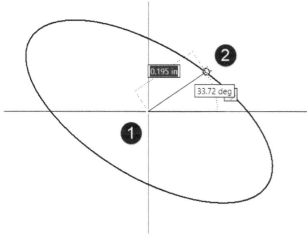

8. Select the centerpoint of the ellipse. Move the pointer toward the right and select a point on the ellipse.

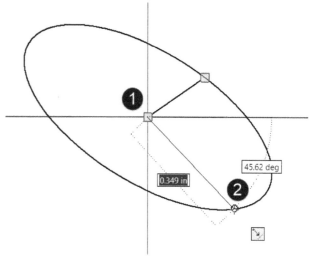

9. On the ribbon, click **Sketch** tab > **Constraint** panel > **Perpendicular**.
10. Select the two lines to make them perpendicular.

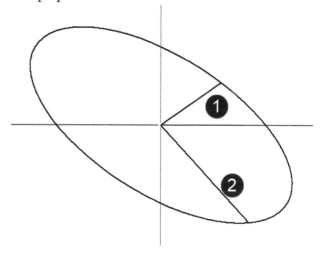

11. On the ribbon, click **Sketch** tab > **Constrain** panel > **Parallel Constraint**.
12. Select the line and the ellipse, as shown.

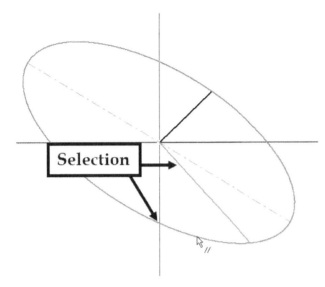

13. Create a horizontal line originating from the center point of the ellipse.

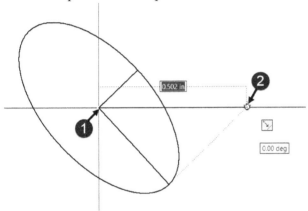

14. Right click and select **OK**.
15. Press and hold the Shift key and select the three lines.
16. On the ribbon, click **Sketch** tab > **Format** panel > **Construction** .
17. On the ribbon, click **Sketch** tab > **Constrain** panel > **Dimension** .
18. Select the lines, as shown.
19. Move the pointer between the selected lines and click to position the angled dimension.

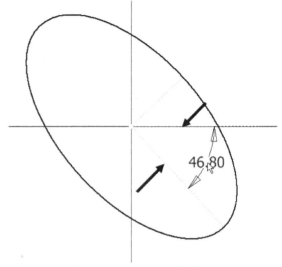

20. Type 15 and click the green check.
21. Select the major axis line.
22. Right click and select **Aligned**.
23. Move the pointer downward and click to position the dimension.
24. Type 0.98 and click the green check.

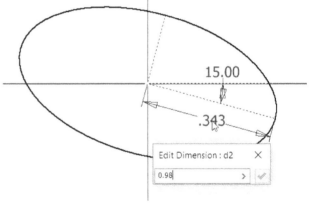

25. Likewise, add the dimension to the minor axis. This fully-defines the sketch.
26. Right-click and select **OK**.

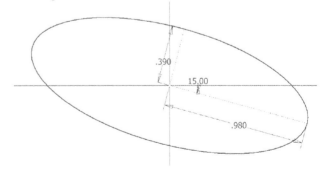

Bridge Curve

The **Bridge Curve** tool is used to create a curve connected two open sketch elements.

1. Create a line and arc, as shown.

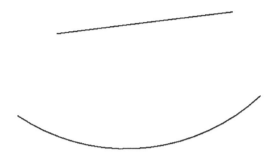

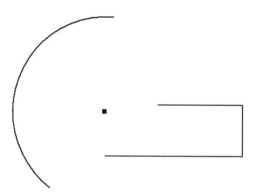

2. On the **Sketch** ribbon, click **Create** panel >

   Line drop-down > **Bridge Curve** .

3. Select the endpoints of the line and arc.

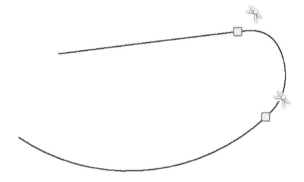

Extend

The **Extend** tool is similar to the **Trim** tool, but its use is opposite of it. This tool is used to extend lines, arcs, and other open entities to connect to other objects.

1. Create a sketch as shown below.

2. Click **Sketch > Modify > Extend** on the ribbon.

3. Select the horizontal open line. This will extend the line up to arc.

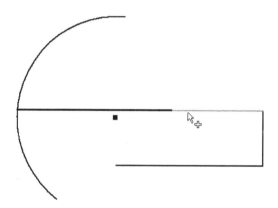

Likewise, extend the other elements, as shown.

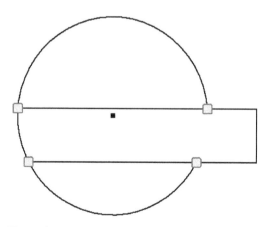

4. Trim the unwanted portions.

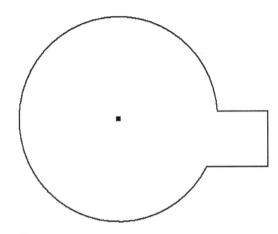

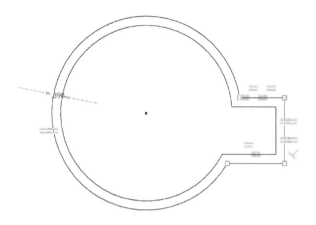

Offset

The **Offset** tool creates parallel copies of lines, circles, arcs, and so on.

1. On the ribbon, click **Sketch** tab > **Modify** panel > **Offset** ⌒.
2. Select an entity and notice that all the connected entities are selected.
3. Type-in the offset distance in the value box attached to the offset copy.
4. Click inside or outside the sketch to define the offset side.

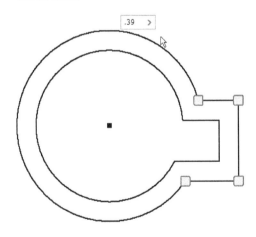

◎ Concentric Constraint

Use the **Concentric Constraint** to make two center points of circles, arcs, or ellipses coincident with each other.

1. Create two circles, as shown.

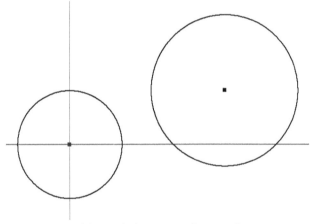

2. On the ribbon, click **Constrain** panel > **Concentric Constraint**.
3. Select the two circles to make them concentric.

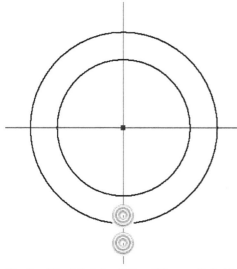

4. On the **Sketch** tab of the ribbon, click **Constrain > Dimension** .

5. Select the inner circle, move the point upward, and click to position the dimension.

6. Type 1.96 and click the green check on the Edit Dimension box.

7. Likewise, add dimension to the outer circle and change its value to 5.5.

8. Right click and select **OK**.

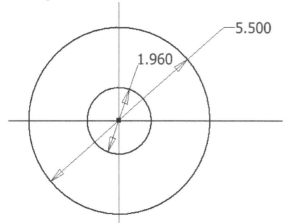

Center Point Arc

The Center Point Arc tool creates an arc by using three points (center, start, and end points).

1. On the ribbon, click **Sketch > Create > Arc drop-down > Arc Center Point**.

2. Select the quadrant point of the circle, as shown below.

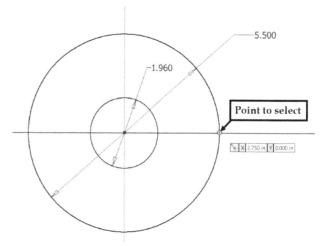

3. Move the pointer down and click on the circle.

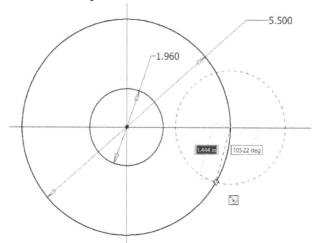

4. Move the pointer leftwards and up, and then click.

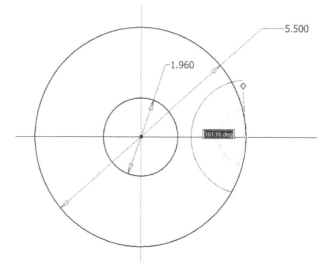

5. Right click and select **OK**.

6. On the ribbon, click **Sketch** tab > **Constrain** panel > **Coincident Constraint** .

7. Select the end point of the arc, and then click on the circle.

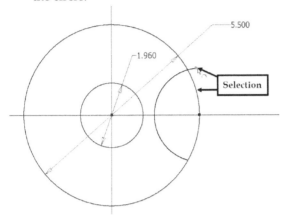

8. On the ribbon, click **Sketch** tab > **Constrain** panel > **Dimension**.
9. Select the arc, move the pointer, and then click to position the dimension.
10. Type 0.98 in the **Edit dimension** box, and then click the green check.

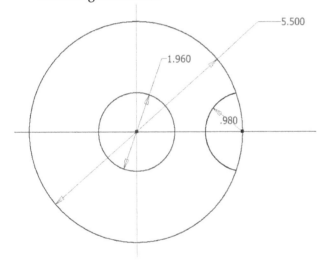

Circular Pattern

The **Circular Pattern** tool creates an arrangement of objects around a point in circular form.

1. On the ribbon, click **Sketch** tab > **Pattern** panel > **Circular Pattern** .
2. Select the arc.
3. Click the cursor button next to the **Axis** button.
4. Select the center of the circle; the preview of the circular pattern appears.

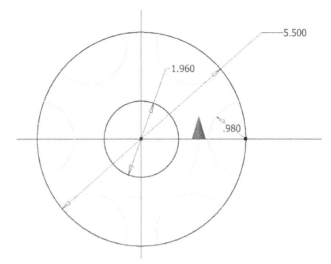

5. Type **4** and **360** in the **Count** and **Angle** boxes, respectively.

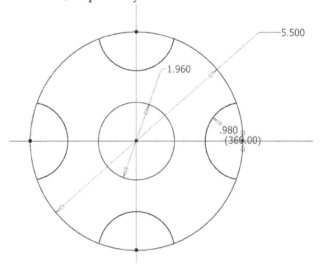

6. Trim the unwanted portions, as shown below.

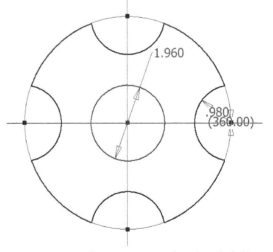

Next, you need to constrain the sketch fully.

7. Click and drag any one of the center points of the arcs. Notice that the arcs are moved freely.
8. On the ribbon, click **Sketch** tab > **Constrain** panel > **Coincident Constraint** .
9. Select the centerpoint of any one of the arcs.
10. Select the outer circle.

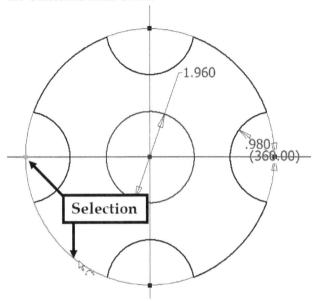

11. Right click and select **OK**.
12. Add dimension to the outer circle.
13. On the ribbon, click **Sketch** tab > **Constrain** panel > **Horizontal Constraint** .
14. Select the center points of the arc and circle, as shown.

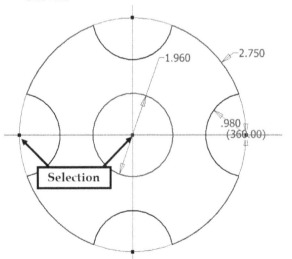

15. Right click and click **OK**.
16. Click **Finish Sketch**.

Text

Texts are used to provide information on the models such as company name, model specifications, and so on.

1. Start a new sketch.
2. Create the circle and two lines perpendicular to each other, as shown.

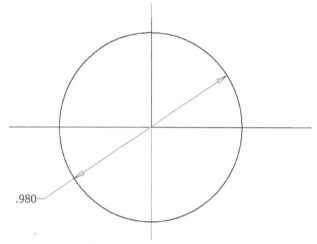

3. Press and hold the Shift key, and then select the circle and the lines.
4. On the ribbon, click **Sketch** tab > **Format** panel > **Construction** .

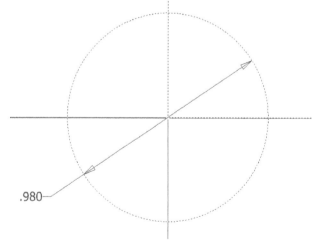

5. On the **Sketch** ribbon, click **Create** panel > **Text** drop-down > **Geometry Text** .
6. Select the horizontal line as the supporting curve.
7. Click in the text box in the Geometry Text dialog, and then type INVENTOR. On the dialog, use the **Position** options (**Left**

Justification 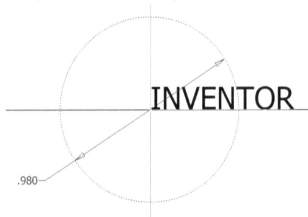, **Center Justification** , **Right Align**) to change the alignment of the text. Use the **Outside** , **Inside** and **Direction options** to the flip the text.

13. Click the **Geometry** button.
14. Select the circle as the alignment curve.
15. Type 0 in the **Start Angle** box.
16. Click **Update**.

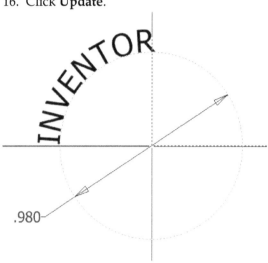

17. Select **Center Justification** from the **Position** group.
18. Type **90** in the **Start Angle** box, and then click **Update**.

8. Click the **Launch Font Editor** icon to display the **Format Text** dialog
9. On the **Format Text** dialog, change the font type, size, font style, and spacing between the text. Also, you can apply effects such as strikeout and underline.
10. Click **OK** on the **Format Text** dialog.
11. Click the **Geometry** button and click on the vertical construction line.
12. Click the **Update** button.

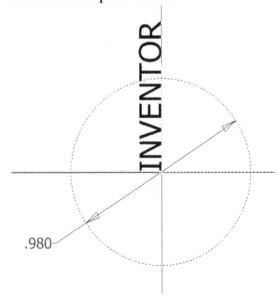

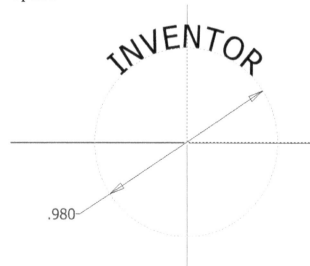

19. Click the **OK** on the **Geometry-Text** box.

Chapter 6: Additional Modeling Tools

In this chapter, you create models using additional modeling tools. You will learn to:

- Create slots
- Create circular patterns
- Create holes
- Create chamfers
- Create shells
- Create rib features
- Create coils
- Create a loft feature
- Create an emboss feature
- Create a thread
- Create a sweep feature
- Create a grill feature
- Create a replace faces
- Create a face fillet
- Create a variable fillet
- Create a boss feature
- Create a lip feature

TUTORIAL 1

In this tutorial, you create the model shown in the figure:

Creating the First Feature

1. Create a new project with the name **Autodesk Inventor 2020 Basics Tutorial** (See Chapter 2, Tutorial 1, Creating New Project section to learn how to create a new project).
2. Open a new Inventor part file using the **Standard.ipt** template (See Chapter 2, Tutorial 3, Starting a New Part File section).
3. Click the **Start 2D sketch** button on the ribbon, and select the XY Plane.

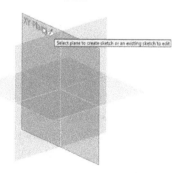

4. Click the **Circle Center Point** button and draw a circle (See Chapter 2, Tutorial 1, Starting

a Sketch section to learn how to create a new project).

5. Click the **Line** button.
6. Specify a point at the top left outside the circle.
7. Move the pointer horizontally and notice the Horizontal constraint symbol.
8. Click outside the circle – Press Esc to deactivate the **Line** tool.

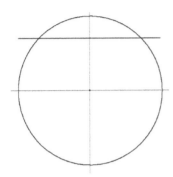

9. Click the **Trim** button on the **Modify** panel.
10. Click on the portions of the sketch to trim, as shown below.

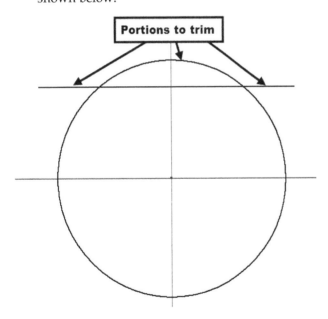

11. Apply dimensions to the sketch (Radius=0.63, vertical length=1.102). To apply the vertical length dimension, activate the **Dimension** tool, and select the horizontal line. Move the pointer downward and place the cursor on the bottom quadrant point of the arc. Click when the

symbol ⌀ appears. Move the pointer toward the left and click to place the dimension.

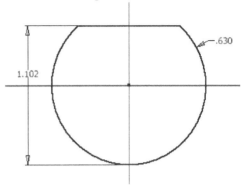

12. Click **Rectangle > Slot Center Point Arc** on the **Create** panel.
13. Select the origin as the center point.
14. Move the cursor outside and click in the first quadrant of the circle to specify the start point of the slot arc.
15. Move the cursor and click in the fourth quadrant of the circle to specify the endpoint of the slot arc.

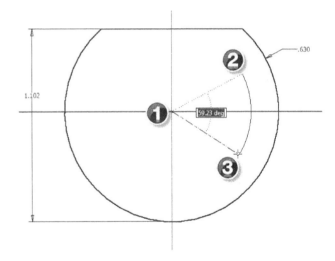

16. Move the cursor outward from the arc and click.

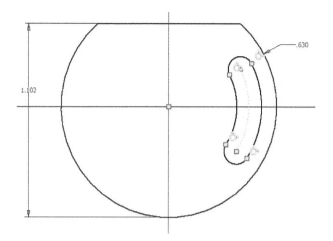

17. Click the **Dimension** button on the **Constrain** panel.
18. Select the start point of the slot arc.
19. Select the center point of the slot arc.
20. Select the end point of the slot arc.

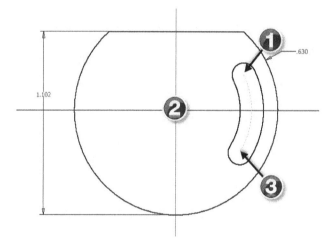

21. Place the angular dimension of the slot; the **Edit Dimension** box appears.
22. Enter **30** in the **Edit Dimension** box and click the green check.

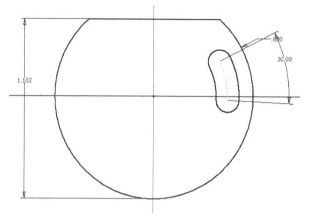

23. Click the **Construction** button on the **Format** panel.

24. Click the **Line** button on the **Create** panel.
25. Draw a horizontal line passing through the origin.

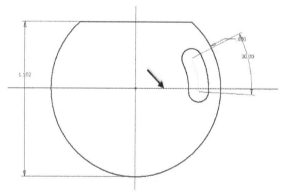

26. Click the **Construction** button on the **Format** panel to deactivate it.
27. Click the **Symmetric** button on the **Constrain** panel.
28. Select the end caps of the slot.
29. Select the construction line; the slot is made symmetric about the construction line.

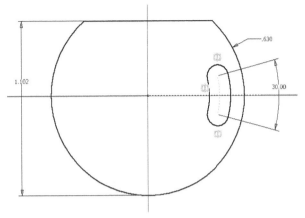

30. Apply other dimensions to the slot, as shown.

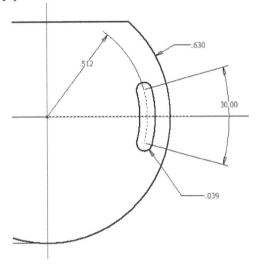

31. Click the **Circular Pattern** button on the **Pattern** panel; the **Circular Pattern** dialog appears.

32. Select all the elements of the slot by creating a selection window.

33. Click the cursor button located on the right-side on the dialog.
34. Select the origin point of the sketch axis point.
35. Enter **4** in the **Count** box and **180** in the **Angle** box.
36. Click the **Flip** button.

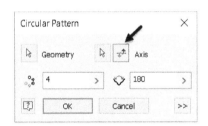

The preview of the circular pattern appears.

37. Click **OK** to create the circular pattern.

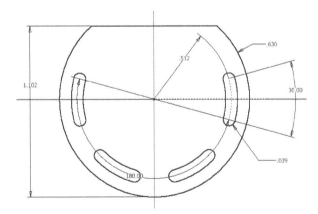

38. Click the **Finish Sketch** button on the ribbon.
39. Extrude the sketch up to 0.236 distance (See Chapter 2, Tutorial 1, Creating the Base Feature section to learn how to create a new project).

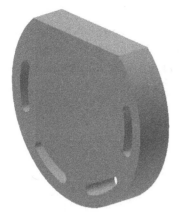

Adding the Second feature

1. Create a sketch on the back face of the model (use the Orbit tool available on the Navigate bar to rotate the model).

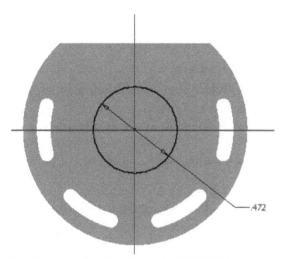

2. Extrude the sketch up to 0.078 distance.

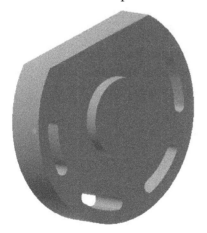

Creating a Counterbore Hole

In this section, you will create a counterbore hole concentric to the cylindrical face.

1. Click the **Hole** button on the **Modify** panel; the **Properties panel** appears.

2. Set the parameters in the **Properties** panel, as shown in the figure.

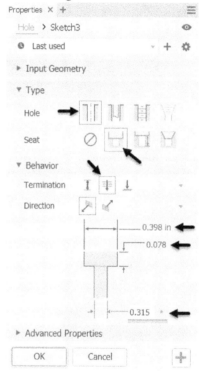

3. Click on the front face of the model; the preview of the hole appears.

Now, you need to specify the reference.

4. Select the cylindrical face of the model; the hole is made concentric to the model.

87

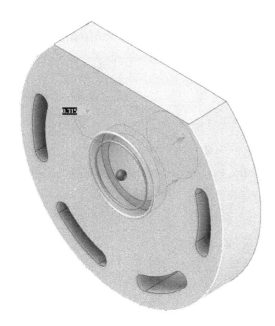

5. Click **OK** on the **Properties** panel; the counterbore hole is created.

Creating a Threaded Hole

In this section, you will create a hole using a sketch point.

1. Click the **Start 2D Sketch** button and select front face of the model.
2. Click the **Point** button on the **Create** panel.

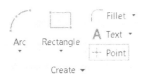

3. Place the point on the front face of the model.
4. Click the **Horizontal** button on the

Constrain panel.
5. Select the point and sketch origin; the point becomes horizontal to the origin.
6. Create a horizontal dimension of 0.354 between point and origin.

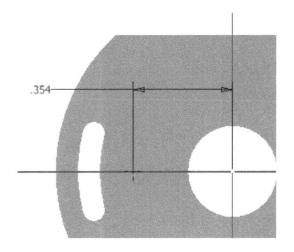

7. Click **Finish Sketch**.
8. Click the **Hole** button on the **Modify** panel; the **Properties** panel appears.
9. Select the sketch point, if not already selected.
10. Select the **Counterbore** option.
11. Select the **Tapped Hole** option.
12. Set the **Thread Type** to **ANSI Unified Screw Threads**.
13. Set the **Size** to **0.073**.
14. Select **Class > 2B**.
15. Set the **Designation** to **1-64 UNC**.
16. Select the **Full Depth** option.
17. Set the **Direction** to **Right Hand**.
18. Set the **Counterbore Diameter** to 0.118.
19. Set the **Counterbore Depth** to 0.039.

20. Click **OK** to create the hole.

Creating a Circular Pattern

1. Click the **Circular Pattern** button on the **Pattern** panel; the **Circular Pattern** dialog appears.
2. Select the threaded hole created in the previous section.
3. Click the **Rotation Axis** button on the dialog.

4. Select the outer cylindrical face of the model.
5. Enter **6** in the **Occurrence** box and **360** in the **Angle** box.
6. Click **OK** to create the circular pattern.

Creating Chamfers

1. Click the **Chamfer** button on the **Modify** panel.
2. Click the **Distance and Angle** button on the dialog.

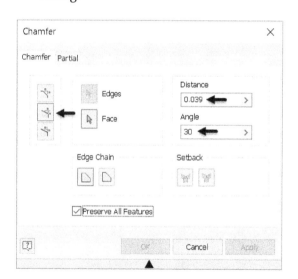

3. Select the cylindrical face of the counterbore hole located at the center.

Face selected

4. Select the circular edge of the counterbore hole.

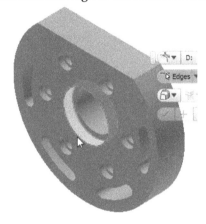

5. Enter 0.039 in the **Distance** box and 30 in the **Angle** box.

6. Click **OK** to create the chamfer.
7. Save the model and close it.

TUTORIAL 2

In this tutorial, you will create the model shown in the figure.

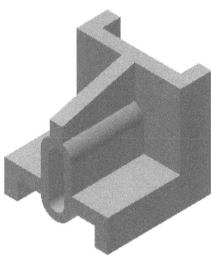

Creating the first feature

1. Open a new Inventor part file using the **Standard.ipt** template (See Chapter 2, Tutorial 3, Starting a New Part File section).
2. On the ribbon, click **3D Model > Sketch > Start 2D Sketch** .
3. Select the YZ plane.
4. Draw an L-shaped sketch using the **Line** tool and dimension it to be 1.575, as shown in the figure.

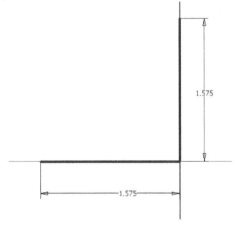

5. Click the **Offset** button on the **Modify** panel.

6. Select the sketch, and then specify the offset position at a random distance.

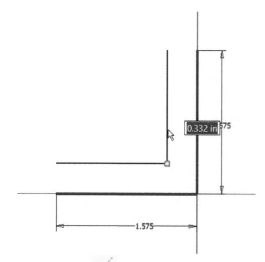

7. Click the **Line** button and draw lines closing the offset sketch.

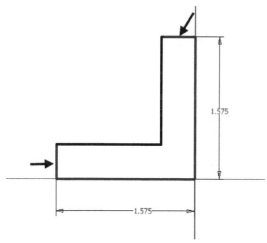

8. Add the offset dimension to the sketch.

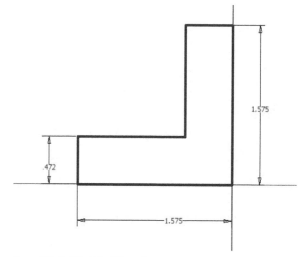

9. Click **Finish Sketch**.

10. Click **3D Model > Create > Extrude** on the ribbon.

11. Select the **Symmetric** option on the **Extrude Properties** panel.

12. Set the **Distance A** to 1.575.

13. Click **OK** to create the first feature.

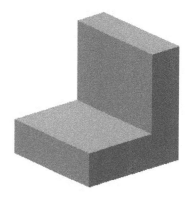

Creating the Shell feature

You can create a shell feature by removing the face of the model and applying thickness to other faces.

1. Click **3D Model > Modify > Shell** on the ribbon; the **Shell** dialog appears. Click the down arrow to expand the dialog
2. Set **Thickness** to 0.197.

Now, you need to select the faces to remove.

3. Select the top face and the back face of the model.

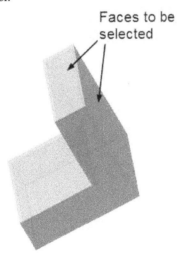

Faces to be selected

4. Select the front face and the bottom face of the model.

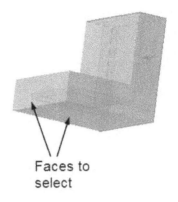

Faces to select

5. Click **OK** to shell the model.

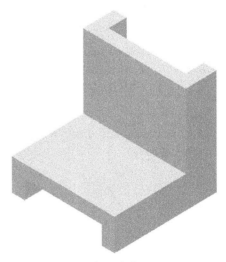

Creating the Third feature

1. Click **3D Model > Sketch > Start 2D Sketch** on the ribbon.

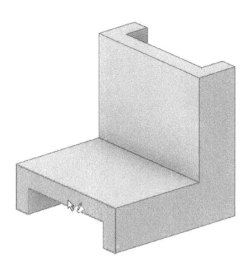

2. Select the front face of the model.
3. Click **Sketch > Create > Rectangle** drop-down **>** **Slot Center to Center** on the ribbon.

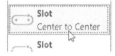

4. Draw a slot by selecting the first, second, and third points. Make sure that the second point is coincident with the lower horizontal edge.

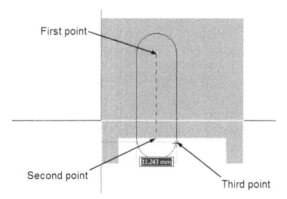

5. Apply dimensions to the slot.

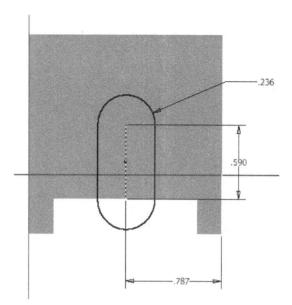

6. Click **Finish Sketch**.
7. Click **3D Model > Create > Extrude** on the ribbon.
8. Select the sketch.
9. Click the **To** button under the **Behavior** section.
10. Select the back face of the model.

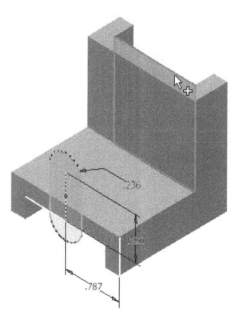

11. Click the **Join** button on the **Extrude Properties** panel.

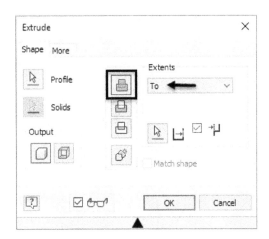

12. Click **OK** to create the feature.

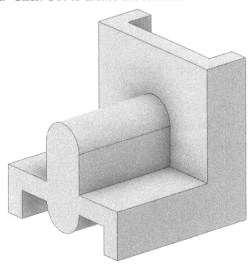

Creating a Cut Feature

1. Create the sketch on the front face of the model, as shown below.

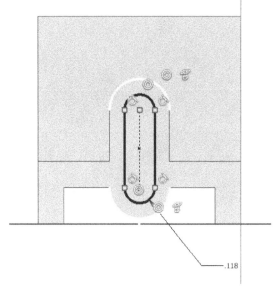

2. Finish the sketch.
3. Click **3D Model > Create > Extrude** on the ribbon.
4. Select the sketch.
5. Click the **Through All** ⬇ button under the **Behavior** section.
6. Click the **Cut** button on the **Extrude Properties** panel.

7. Click **OK** to create the cut feature.

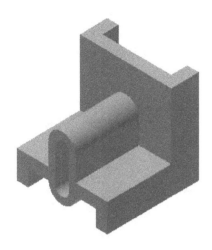

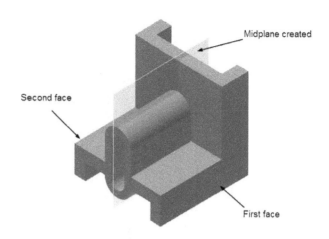

Creating the Rib Feature

In this section, you will create a rib feature in the middle of the model. To do this, you must create a mid-plane.

1. To create a mid-plane, click **3D Model > Work Features > Plane > Midplane between Two Planes** on the ribbon.

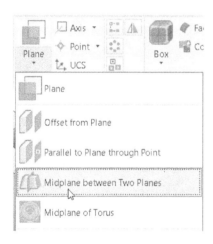

2. Select the right face of the model.
3. Select the left face of the model; the midplane is created.

4. Click **3D Model > Sketch > Start 2D Sketch** on the ribbon.
5. Select the midplane.
6. Click the **Slice Graphics** button at the bottom of the window.

7. Click **Sketch > Create > Project Geometry > Project Cut Edges** on the ribbon; the edges cut by the sketch plane are projected.

8. Draw the sketch, as shown below.

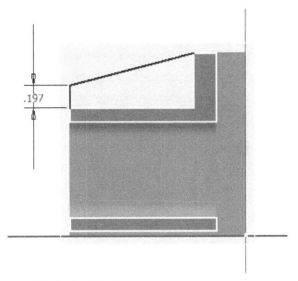

9. Finish the sketch.

10. Click **3D Model > Create > Rib** on the ribbon; the **Rib** dialog appears.

11. Select the sketch.

12. Click the **Parallel to Sketch Plane** button on the dialog.

13. Click the **Direction 1** button.

14. Set **Thickness** to 0.197.

15. Click the **To Next** button.

16. Click the **Symmetric** button below the **Thickness** box.

17. Click **OK** to create the rib feature.

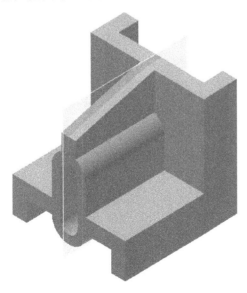

18. To hide the midplane, select it and right-click.

19. Click **Visibility** on the Marking Menu; the plane will be hidden.

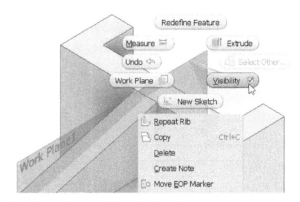

20. Save the model and close it.

TUTORIAL 3

In this tutorial, you will create a helical spring using the **Coil** tool.

Creating the Coil

1. Open a new Inventor file using the **Standard.ipt** template (See Chapter 2, Tutorial 3, Starting a New Part File section).

2. Click the **Start 2D sketch** button on the ribbon, and select the XY Plane.

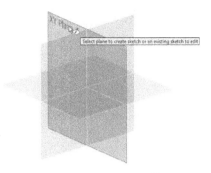

3. Click the **Circle Center Point** button.
4. Select a point on the left portion of the horizontal axis
5. Move the cursor outward and click to create a circle.
6. On the ribbon, click **Sketch > Format > Centerline** .
7. Click the **Line** button.
8. Select the origin point of the sketch, move the cursor upward, and click to create a centreline.
9. On the ribbon, click **Sketch > Constrain > Dimension**.
10. Add the diameter dimension to the circle, as shown.
11. Select the centreline and circle.
12. Move the pointer downward and click.
13. Type 1.575 and press Enter.

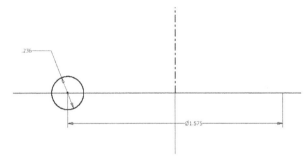

14. Finish the sketch.
15. To create a coil, click **3D Model > Create > Coil** on the ribbon; the **Coil** dialog appears.

In addition, the profile is automatically selected. Now, you need to select the axis of the coil.

16. Select the centerline as the axis. Click the

Reverse Axis button on the **Coil** dialog, if the coil preview is downwards.

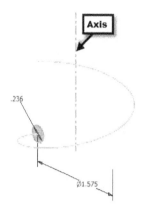

17. Click the **Coil Size** tab on the dialog.
18. In the **Coil Size** tab, specify the settings as given next.

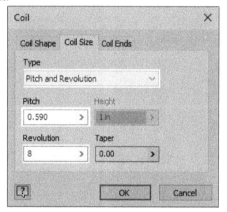

19. Click the **Coil Ends** tab on the dialog.
20. Specify the settings in the **Coil Ends** tab, as given next.

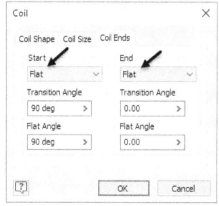

21. Click **OK** to create the coil.

22. Save the model as **Coil.ipt** and close the file.

TUTORIAL 4

In this tutorial, you create a shampoo bottle using the **Loft**, **Extrude**, and **Coil** tools.

Creating the First Section and Rails

To create a swept feature, you need to create sections and guide curves.

1. Open a new Part file (See Chapter 2, Tutorial 3, Starting a New Part File section).
2. Click **3D Model > Sketch > Start 2D Sketch** on the ribbon.

3. Select the XZ plane.

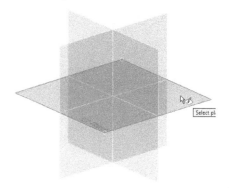

4. Click **Sketch > Create > Circle > Ellipse** on the ribbon.
5. Draw the ellipse by selecting the points, as shown.

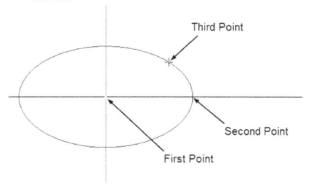

6. On the ribbon, click **Sketch > Constrain > Dimension**.
7. Select the ellipse, move the cursor downward, and click.
8. Type 1.968 and press Enter.
9. Select the ellipse, move the cursor toward left, and click.
10. Type 0.984 and press Enter.

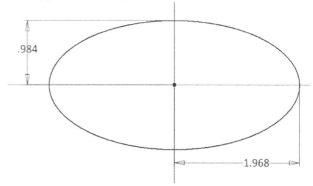

11. Click **Finish Sketch**.

Additional Modeling Tools

12. Click **3D Model > Sketch > Start 2D Sketch** on the ribbon.
13. Select the YZ plane from the graphics window.
14. Click **Sketch > Create > Line> Spline Interpolation** on the ribbon.

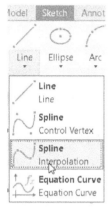

15. Select a point on the horizontal axis of the sketch; a rubber band curve is attached to the cursor

16. Move the cursor up and specify the second point of the spline; a curve is attached to the cursor.
17. Move the cursor up and specify the third point.
18. Likewise, specify the other points of the spline, as shown.
19. Click the right mouse button and select **Create**. The spline will be similar to the one shown in the figure.

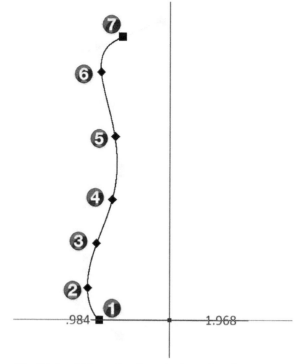

20. Right click and select **Create Line** from the Marking Menu.
21. On the ribbon, click **Sketch > Format > Construction** .
22. Select the origin point of the sketch, move the pointer vertically upward and click to create a vertical construction line.

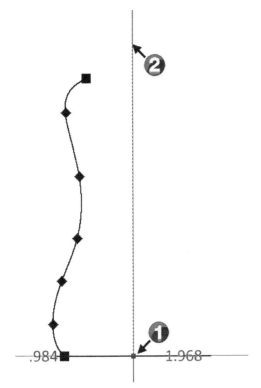

23. On the ribbon, click **Sketch > Constrain > Vertical** .
24. Select the lower end points of the vertical construction line and the spline.

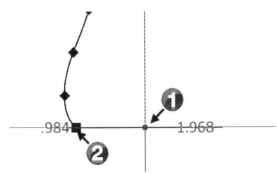

25. On the ribbon, click **Sketch > Constrain > Dimension**.
26. Select the lower end points of the construction line and spline.
27. Move the cursor downward and click.
28. Type the 1.968 in the Edit Dimension box and press Enter.

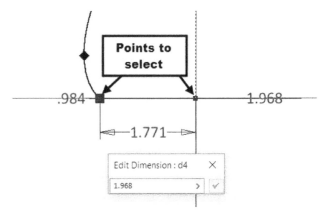

29. Select the second point of the spline and the construction line.
30. Move the cursor and click to place the dimension.
31. Type 2.362 in the **Edit Dimension** box and press Enter.

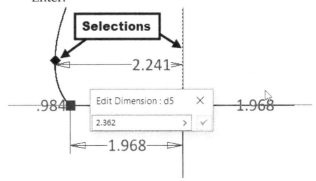

32. Apply the other horizontal dimensions to the spline, as shown in the figure.

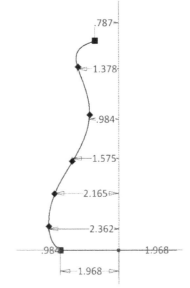

33. Select the origin point of the sketch and the top endpoint of the spline.
34. Move the pointer toward the right and click.

35. Type 8.858 in the **Edit Dimension** box and press Enter.

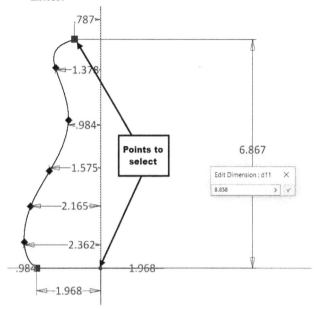

36. Likewise, create other dimensions, as shown.

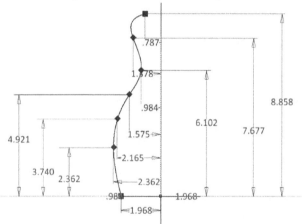

37. Click **Sketch > Pattern > Mirror** on the ribbon; the **Mirror** dialog appears.
38. Select the spline. Make sure that you select the curve and not the points.
39. Click **Mirror line** on the **Mirror** dialog, and then select the construction line.
40. Click **Apply**, and then click **Done**.

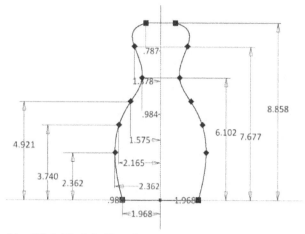

41. Click **Finish Sketch**.

Creating the second section

1. Click **3D Model > Work Features > Plane > Offset from Plane** on the ribbon.

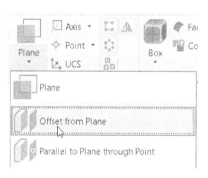

2. Select the XZ plane from the Browser window.

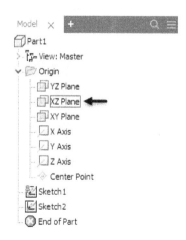

3. Enter **8.858** in the **Distance** box.

101

4. Click **OK** ✓.
5. On the ribbon, click the **3D Model > Sketch > Start 2D Sketch**.
6. Select the newly created datum plane.
7. Right click and select **Center Point Circle** from the Marking Menu.
8. Select the origin point, move the cursor outside, and click to create a circle.
9. Right click and select **OK**.
10. Right click and select **General Dimension** from the Marking Menu.
11. Select the circle, move the cursor outward, and click.
12. Type 1.574 and press Enter.

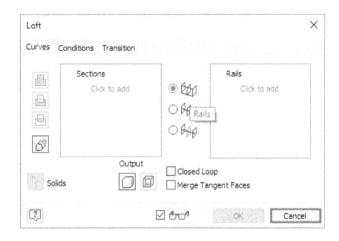

3. Click **Click to add** in the **Sections** group and select the circle.

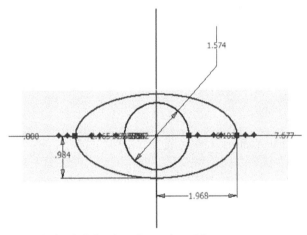

13. Click **Finish Sketch** on the ribbon.

Creating the Loft feature

1. To create a loft feature, click **3D Model > Create > Loft** ⌬ on the ribbon; the **Loft** dialog appears.
2. Select the **Rails** option from the dialog.

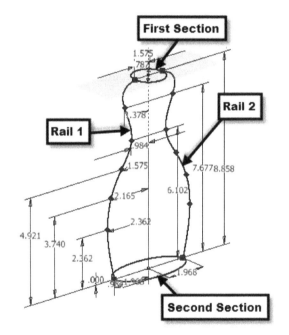

4. Select the ellipse.
5. Click **Click to add** in the **Rails** group.
6. Select the first rail.
7. Select the second rail.
8. Click **OK** to create the loft feature.

Creating the Extruded feature

1. Right click and select **New Sketch** from the Marking Menu.
2. Select the plane located at the top of the sweep feature.
3. Right click and select **Center Point Circle** from the Marking Menu.
4. Select the origin point, move the cursor outside, and click to create a circle.
5. Right click and select **OK**.
6. Right click and select **General Dimension** from the Marking Menu.
7. Select the circle, move the cursor outward, and click.
8. Type 1.574 and press Enter.
9. Click **Finish Sketch** on the ribbon.
10. Click the **Extrude** button on the **Create** panel.
11. Extrude the circle up to 1 in.

Creating the Emboss feature

1. Click **3D Model > Work Features > Plane > Offset from Plane** on the ribbon.
2. Select the YZ plane from the Browser window.

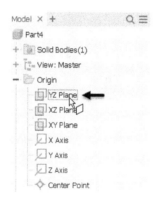

3. Enter **2** in the **Distance** box and click **OK**.

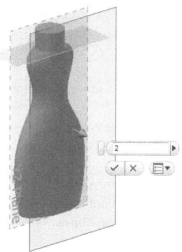

4. Click **3D Model > Sketch > Start 2D Sketch** on the ribbon.
5. Select the newly created datum plane
6. Click **Sketch > Create > Circle > Ellipse** on the ribbon.
7. Draw the ellipse by selecting the points, as shown.

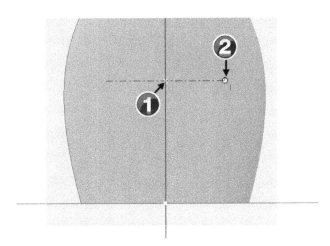

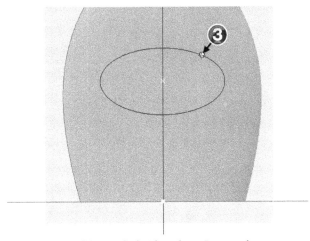

8. On the ribbon, click **Sketch > Constrain > Dimension**.
9. Select the ellipse, move the cursor downward, and click
10. Type 2 and press Enter.
11. Select the ellipse, move the cursor toward left, and click.
12. Type 1.35 and press Enter.
13. Select the origin point of the sketch and the center point of the ellipse.

14. Move the cursor toward left and click to place the dimension between the selected point.
15. Type 2.55 and press Enter.
16. Right click and select **Line** from the Marking Menu.
17. Right click and select **Construction** from the Marking Menu.
18. Select the origin point of the sketch and the center point of the ellipse.
19. Right click and select **OK**; a construction line is created between the sketch origin and the ellipse. Also, the sketch is fully-constrained.
20. Click **Finish Sketch**.

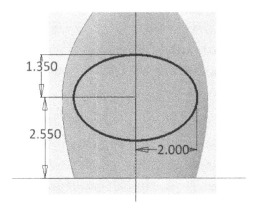

21. Click **3D Model > Create > Emboss** on the ribbon; the **Emboss** dialog appears.
22. Select the sketch, if not already selected.
23. Click the **Engrave from Face** button on the dialog.
24. Set the **Depth** to 0.125.
25. Click **OK** to create the embossed feature.

Mirroring the Emboss feature

1. Click **3D Model > Pattern > Mirror** on the ribbon.
2. Select the emboss feature from the model geometry.
3. On the **Mirror** dialog, click the **Mirror Plane** button, and then select the YZ Plane from the Browser window.
4. Click **OK** to mirror the emboss feature.

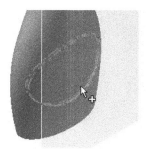

Creating Fillets

1. Click **3D Model > Modify > Fillet** 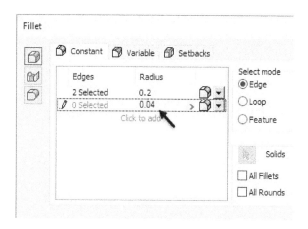 on the ribbon; the **Fillet** dialog appears.
2. Click on the bottom and top edges of the swept feature.
3. Set **Radius** to 0.2.

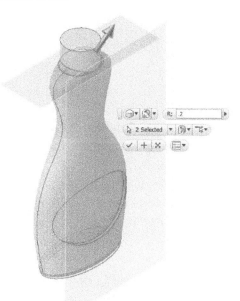

4. Click **Click to add** on the Fillet dialog.
5. Set **Radius** to 0.04.

6. Select the edges of the emboss features, and click **OK**.

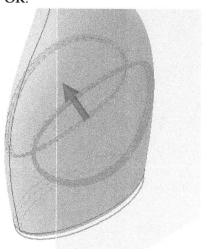

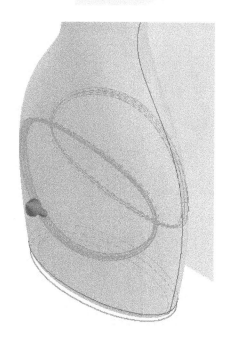

Shelling the Model

1. Click **3D Model > Modify > Shell** on the ribbon; the **Shell** dialog appears.
2. Set **Thickness** to 0.03.
3. Select the top face of the cylindrical feature.

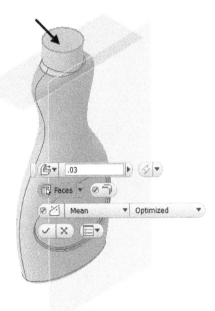

4. Click **OK** to create the shell.

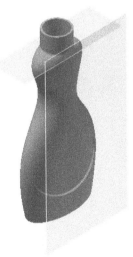

Adding Threads

1. Select the YZ Plane, and then click **Create Sketch**.

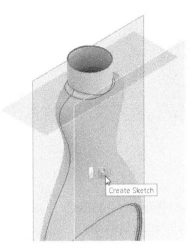

2. Click the **Slice Graphics** icon located at the bottom of the window.

3. On the ribbon, click **Sketch > Create > Line**.
4. Right click and select **Centerline** from the Marking Menu.
5. Select the origin point of the sketch, move the cursor vertically upward and click to create a vertical centreline. Press Esc.
6. Deactivate the **Centerline** icon on the **Format** panel.
7. Right click and select **Create Line** from the Marking Menu.
8. Right click and select **Construction** from the Marking Menu.
9. Create a horizontal construction line, as shown.

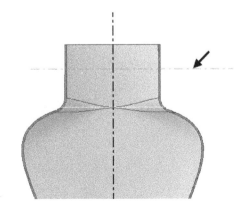

10. Deactivate the **Construction** icon on the **Format** panel of the **Sketch** ribbon tab.
11. On the ribbon, click **Sketch > Create > Line**.
12. Create a closed profile, as shown.

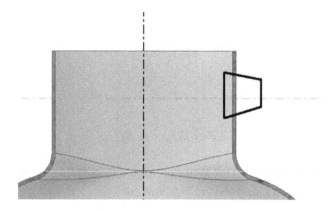

13. On the ribbon, click **Sketch > Constrain > Symmetric** .
14. Select the two inclined lines of the sketch, and then select the construction line; the two inclined lines are made symmetric about the construction line.

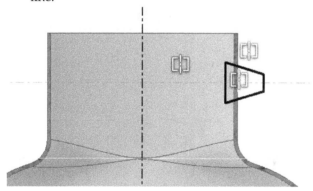

15. Click the **Dimension** button on the **Constrain** panel.
16. Select the two inclined lines, move the cursor horizontally toward left and click.
17. Type 60 in the **Edit Dimension** box and press Enter.

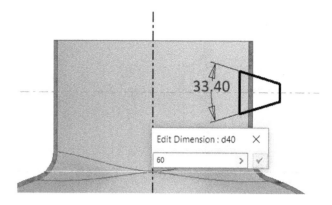

18. Draw the thread profile.

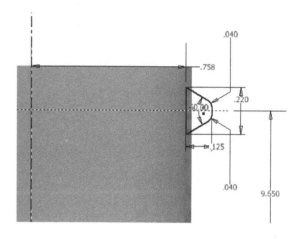

19. On the ribbon, click **Sketch > Create > Fillet** .
20. Type 0.04 in the **2D Fillet** box.
21. Make sure that the **Equal** button is active on the **2D Fillet** box.
22. Select the two corners of the sketch, as shown.

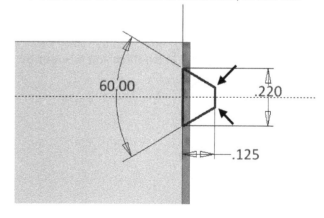

23. Close the **2D Fillet** dialog box.

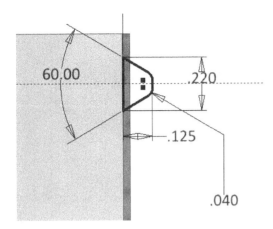

24. Click **Finish Sketch** on the ribbon.
25. On the ribbon, click **3D Model > Create > Coil**
    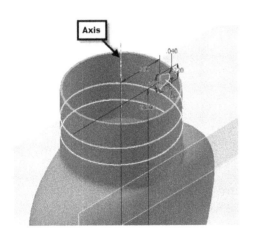 ; the closed profile of the sketch is selected, automatically.
26. Select the axis of the coil.

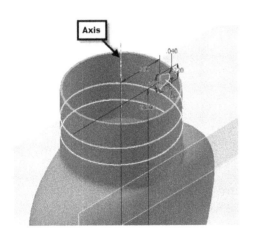

27. On the dialog, click the **Coil Size** tab and select **Type > Pitch and Revolution**.
28. Type-in **0.275** and **2** in the **Pitch** and **Revolution** boxes, respectively.

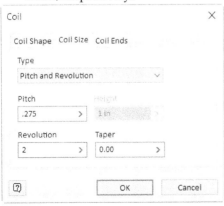

29. Click **OK**.

30. In the Browser Window, right click on the YZ Plane, and then select **New Sketch**.
31. On the ribbon, click **Sketch > Create > Project Cut Edges > Project Geometry** 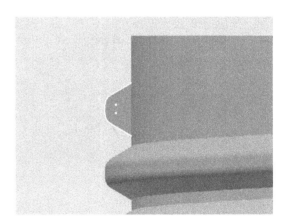 .
32. Select the edges of the end face of the thread.

33. Draw a straight line connecting the endpoints of the projected elements.

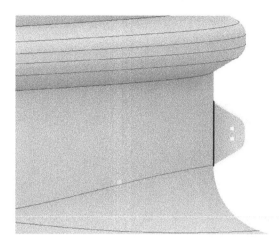

34. Click **Finish Sketch**.
35. Activate the **Revolve** tool and click on the vertical line of the sketch.
36. On the dialog, select **Extents > Angle**, and then type in **100** in the **Angle1** box.
37. Click the **Direction 1** button.

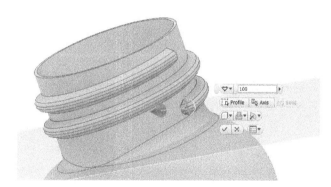

38. Click **OK**.

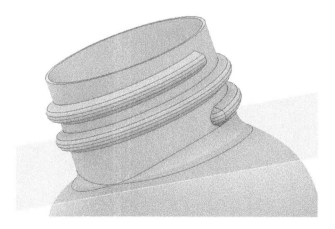

39. Likewise, blend the other end of the thread.

Note that you need the reverse the direction of revolution.

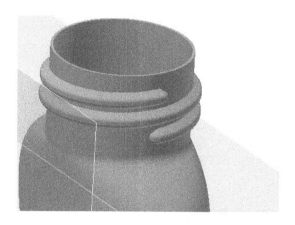

40. Save the model.

TUTORIAL 5

In this tutorial, you create a chair, as shown.

Creating a 3D Sketch

1. Open a new Inventor file using the Standard.ipt template (See Chapter 2, Tutorial 3, Starting a New Part File section).
2. Click the **Home** icon located above the ViewCube. This changes the view orientation to Home.

3. On the ribbon, click **3D Model > Sketch > Start**

2D Sketch > Start 3D Sketch.

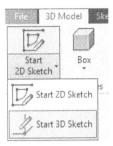

4. On the **3D Sketch** tab of the ribbon, click **Draw > Line**.

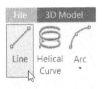

5. Expand the **Draw** panel on the ribbon and activate the **Precise Input** option.

6. On the **Precise Input** toolbar, click the **Reset to Origin** button.
7. Select the **Relative** option from the drop-down available on the **Precise Input** toolbar.

8. On the **Precise Input** toolbar, click in the X box and type 0.
9. Press the Tab key and type 0 in the Y box.
10. Press the Tab key and type 0 in the Z box.
11. Press Enter to specify the first point.

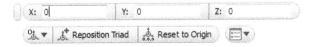

12. Type-in 12 in the **X** box and press Tab on your keyboard.
13. Likewise, type-in 0 in the **Y** and **Z** boxes, respectively. Press Enter to specify the second point.

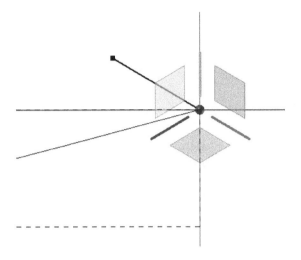

14. Type-in 0, 0, and 20 in the X, Y, and Z boxes, respectively. Press Enter.

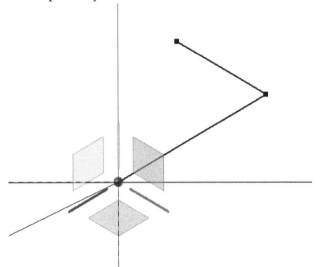

15. Type-in 0, 18, 0 in the X, Y, and Z boxes, respectively. Press Enter.

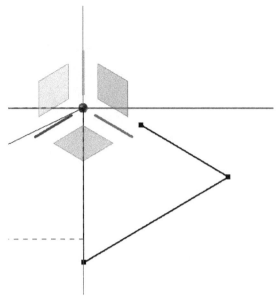

16. Type-in 0, 0 and -22 in the X, Y and Z boxes, respectively. Press Enter.

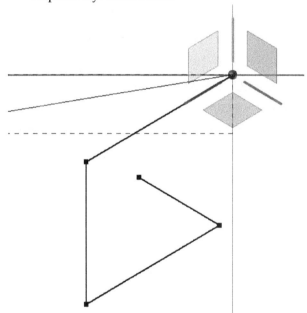

17. Type-in 0, 18, and 0 in the X, Y, and Z boxes, respectively. Press Enter.

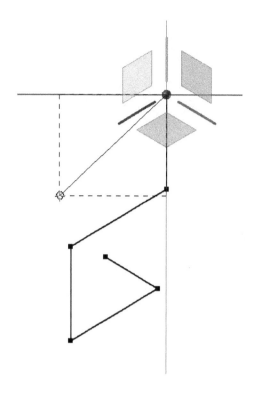

18. Type-in -12, 0, and 0 in the X, Y, and Z boxes, respectively. Press Enter.

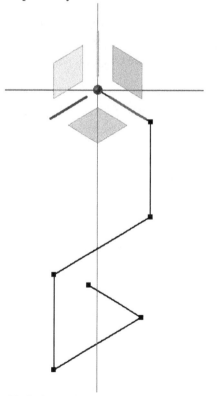

19. Click the right mouse button and select **OK**.
20. On the **3D Sketch** tab of the ribbon, click **Pattern > Mirror**.

111

21. Drag a selection box and select all the sketch elements.

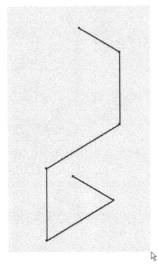

22. Click the **Mirror Plane** button on the dialog, and then select YZ Plane from the Browser window.

23. Click **Apply** and **Done** on the dialog.

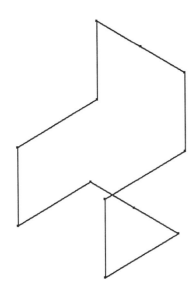

24. On the ribbon, click **3D Sketch > Draw > Bend**.

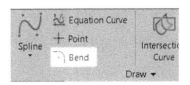

25. Type-in 3 in the **Bend** dialog and select the intersecting lines, as shown in the figure.

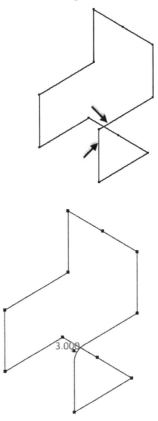

26. Likewise, bend the other corners of the 3D sketch.

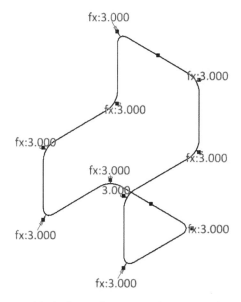

27. Click the right mouse button and select **OK**.
28. On the ribbon, click **3D Sketch > Constrain > Fix** 🔒.
29. Select the origin point of the sketch.

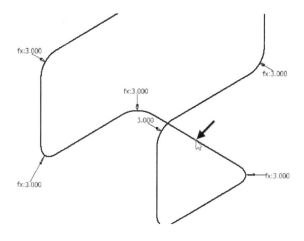

30. Add dimensions to define the sketch fully.

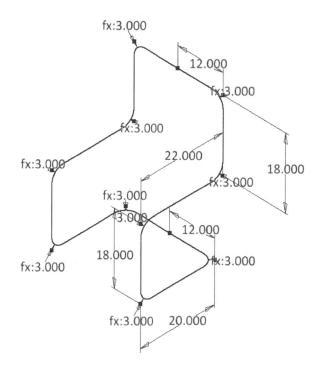

31. Click **Finish Sketch** on the ribbon.
32. On the ribbon, click **3D Model > Work Features > Plane > Normal to Axis through Point**.

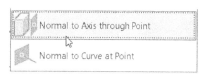

33. Click on the upper horizontal line of the sketch and its endpoint.

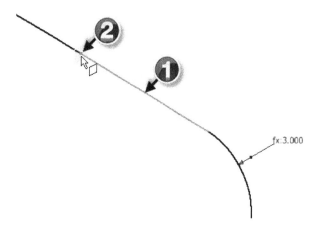

34. Click **3D Model > Sketch > Start 2D Sketch** on the ribbon.

113

35. Select the newly created plane.
36. Create two concentric circles of 1.250 and 1 diameters.

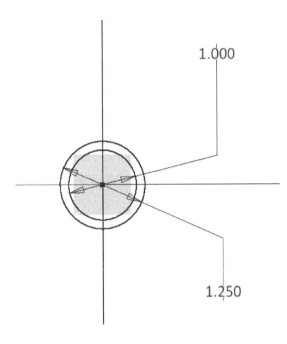

37. Add dimensions from the sketch origin to position the circles.

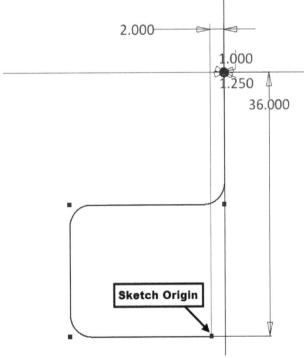

38. Click **Finish Sketch**.

Creating the Sweep feature

1. On the ribbon, click **3D Model > Create > Sweep**.

2. Zoom into the circular sketch and click in the outer loop.

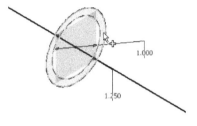

3. Click on the 3D sketch to define the sweep path.

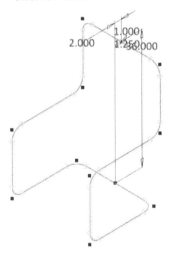

4. Click **OK** to sweep the profile.

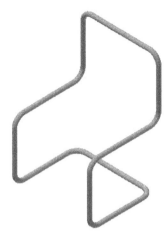

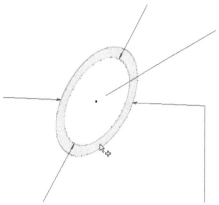

5. Start a sketch on the YZ Plane.
6. Click the **Slice Graphics** icon at the bottom of the graphics window.
7. Draw two concentric circles and dimension them.

10. On the **Extrude Properties** panel, click the **To** ⊥ icon under the **Behavior** section.
11. Click on the tube, as shown.

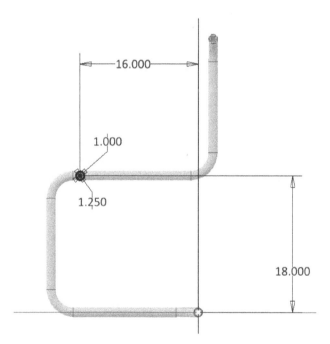

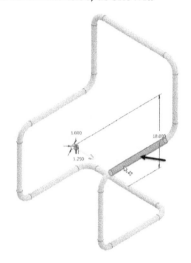

8. Click **Finish Sketch.**
9. Activate the **Extrude** tool and click in the outer loop of the sketch.

12. Click in the **From** selection box in the **Input Geometry** section.
13. Click on the other side of the tube, as shown.

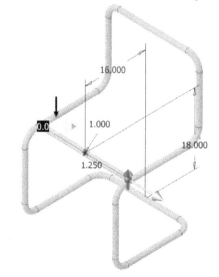

115

14. Make sure that the **Minimum Solution** option active.
15. Click **OK** to extrude the sketch.

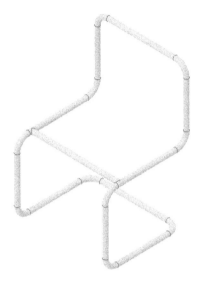

Creating the Along Curve pattern

1. On the Browser window, click the right mouse button on the **3D Sketch** and select **Visibility**; the 3D sketch is displayed.

2. On the ribbon, click **3D Model > Pattern > Rectangular Pattern**.

3. Click on the extrude feature.
4. On the **Rectangular Pattern** dialog, click the **Direction 1** button, and then click on the **3D sketch**.

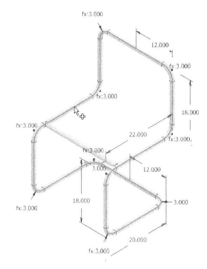

5. Type-in **3** and **23** in the **Column Count** and **Column Spacing** boxes, respectively.
6. Select **Spacing > Distance** from the drop-down menu.
7. Click the double-arrow button located at the bottom of the dialog. This expands the dialog.
8. Set the **Orientation** to **Direction 1**.

9. Click **OK** to pattern the extruded feature.
10. On the Browser window, click the right mouse button on the 3D sketch and select **Visibility**. This hides the 3D sketch.

116

Creating the Freeform feature

1. On the ribbon, click **3D Model > Work Features > Plane**.

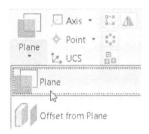

2. On the **Browser window**, click the XZ Plane. A plane appears on the XZ Plane.
3. Click on the top portion of the extruded feature. A plane appears tangent to the extruded feature.

4. Start a sketch on the new plane.
5. Place a point on the sketch plane and add dimensions to position it.

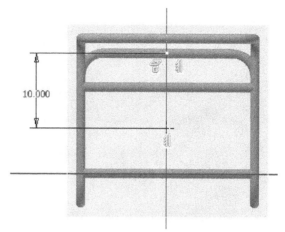

6. Click **Finish Sketch**.
7. On the ribbon, click **3D Model > Create Freeform > Box** 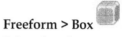 .
8. Select the plane tangent to the extruded feature.
9. Select the sketch point to define the location of the freeform box.
10. Click and drag the side-arrow of the freeform box.

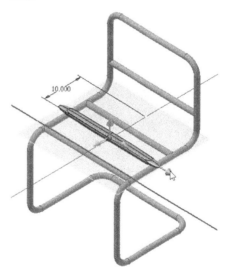

11. Click and drag the front arrow of the freeform box.

117

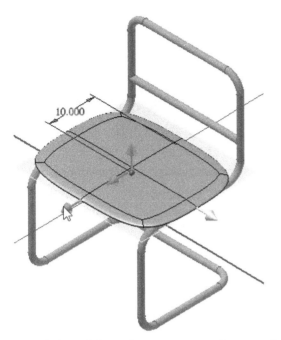

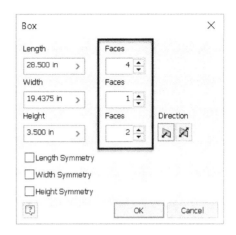

14. Click **OK** to create the freeform shape.

12. Click and drag the top arrow to increase the height of the freeform box.

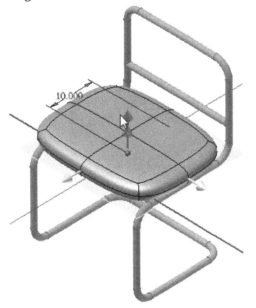

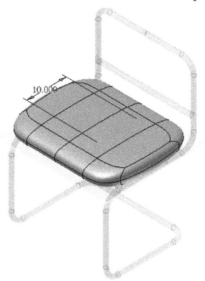

Editing the Freeform Shape

1. On the ribbon, click **Freeform > Edit > Edit Form** .

2. Hold the Ctrl key and click on top faces of the freeform shape.

13. On the **Box** dialog, type in 4, 1, and 2 in the **Faces** boxes, respectively.

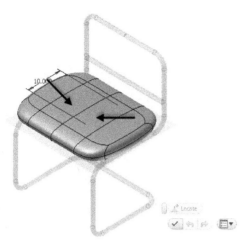

3. Click on the arrow pointing upwards.
4. Drag it downwards 1.6875 in as shown in the figure below.

5. Click **OK** on the dialog.
6. On the ribbon, click **Freeform > Edit > Edit Form** .
7. Hold the Ctrl key and click on the two edges at the front.

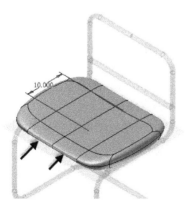

8. Drag the vertical arrow downwards.

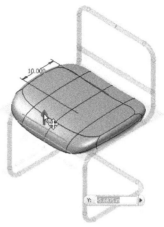

9. Click **OK** on the **Edit Form** dialog.
10. Click **Finish Freeform** on the ribbon.

Create another Freeform box

1. On the ribbon, click **3D Model > Work Features > Plane** .
2. On the Browser window, click the XY Plane.
3. Click on the vertical portion of the sweep feature to create a plane tangent to it.

119

4. Start a sketch on the new plane.
5. Place a point and add dimensions to it.

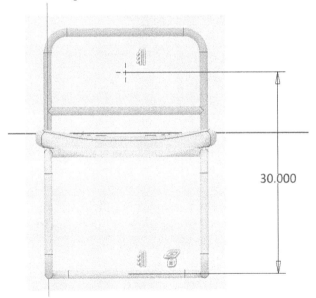

6. Click **Finish Sketch**.

7. Activate the Freeform **Box** tool.
8. Select the new plane and click on the sketch point.
9. On the **Box** dialog, type in 27, 16, and 3 in the

Length, **Width**, and **Height** boxes, respectively.
10. Click **OK**, and then click **Finish Freeform**.

11. Save and close the file.

TUTORIAL 6

In this tutorial, you create a bolt.

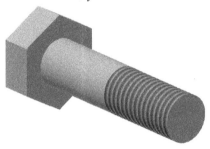

Start a new part file

1. Start a new part file using the **Standard.ipt** template (See Chapter 2, Tutorial 3, Starting a New Part File section).
2. On the ribbon, click **3D Model > Primitives > Primitive drop-down > Cylinder**.
3. Click on the YZ Plane.
4. Click the origin point of the sketch to define the center point of the circle.
5. Move the pointer and type in 0.75 in the box, and then press Enter.

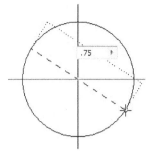

6. Type-in 3 in the **Distance A** box and press Enter.

Creating the second feature

1. Start a sketch on the YZ Plane.

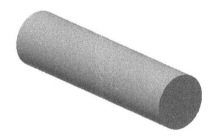

2. On the ribbon, click **Sketch > Create > Rectangle** drop-down **> Polygon**.

3. Click the sketch origin.

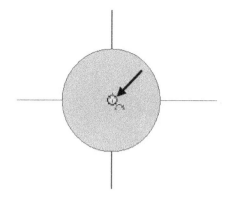

4. Type-in 6 in the **Polygon** dialog.

5. Move the pointer vertically upward. You will notice that a dotted trace line appears between the origin point and the pointer.

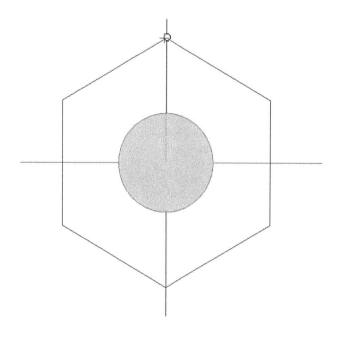

6. Click to create the polygon.
7. Click **Done** on the dialog.
8. On the ribbon, click the **Sketch > Format > Construction** .
9. Activate the **Line** tool and select the vertices of the polygon.

121

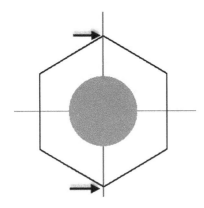

10. Activate the **Dimension** tool and create a dimension, as in the figure.

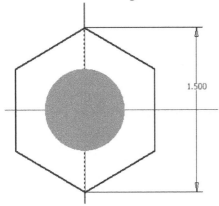

11. Finish the sketch.
12. Activate the **Extrude** tool and select the sketch, if not already selected.
13. On the **Extrude Properties** panel, type-in 0.5 in the **Distance A** box.
14. Use the **Default** or **Flipped** buttons to make sure that the polygon is extruded toward left.
15. Click **OK**.

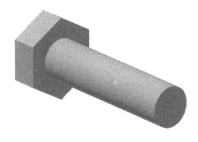

Adding Threads

1. On the ribbon, click **3D Model > Modify > Thread** .
2. Click on the round face of the model geometry.
3. On the **Thread Properties** panel, deactivate the **Full Depth** icon under the **Behavior** section.
4. Type-in **1.5** in the **Depth** field and type in **0** in the **Offset** field.
5. Specify the thread settings under the **Threads** section on the **Thread Properties** panel, as shown in the figure.

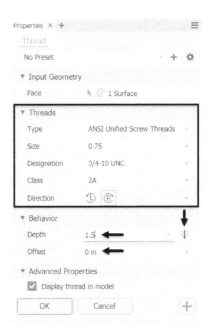

6. Click **OK** to add the thread.

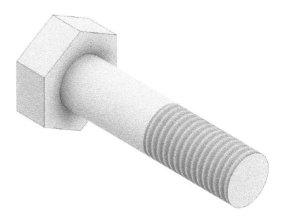

Creating iParts

The iParts feature allows you to design a part with different variations, sizes, materials, and other attributes. Now, you will create different variations of the bolt created in the previous section.

1. On the ribbon, click **Manage > Parameters > Parameters**. This opens the **Parameters** dialog.

2. On the **Parameters** dialog, click in the first cell of the **Model Parameters** table, and type-in **Diameter**.

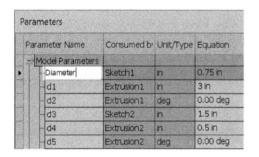

3. Likewise, change the names of other parameters.

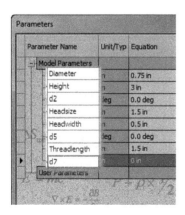

4. Click **Done** on the **Parameters** dialog.
5. Click the right mouse button in the graphics window and select **Dimension Display > Expression**.

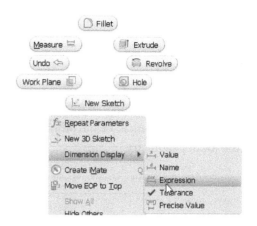

6. On the Browser window, click the right mouse button on the **Extrusion1** and select **Show Dimensions**. You will notice that the dimensions are shown along with the names.

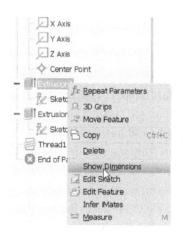

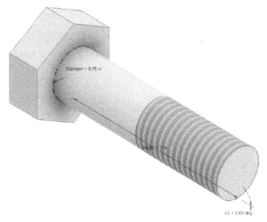

7. On the ribbon, click **Manage > Author > Create iPart**.

This opens the **iPart Author** dialog. In this dialog, you will define the parameters to create other versions of the model geometry.

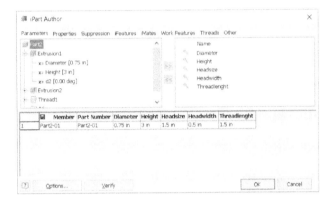

The table at the bottom of this dialog shows the parameters for the iPart factory. You will notice that the renamed parameters are automatically added to the table. If you want to add more parameters to the table, then select them from the section located at the left side. Click the arrow button pointing towards the right. Likewise, if you want to remove a parameter from the table, then select it from the right-side section and click the arrow pointing toward left.

8. Now, click the right mouse button on the table and select **Insert Row**. Notice that a new row is added to the table.

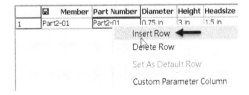

9. Likewise, insert another row.

		Member	Part Number	Diameter	Height	Headsize	Headwidth	Threadlenght
1		Part2-01	Part2-01	0.75 in	3 in	1.5 in	0.5 in	1.5 in
2		Part2-03	Part2-03	0.75 in	3 in	1.5 in	0.5 in	1.5 in
3		Part2-02	Part2-02	0.75 in	3 in	1.5 in	0.5 in	1.5 in

10. In the second row of the table, type-in new values (5, 0.75, and 3) in the Height, Headwidth, and Threadlength boxes. This creates the second version of the bolt.

		Member	Part Number	Diameter	Height	Headsize	Headwidth	Threadlenght
1		Part2-01	Part2-01	0.75 in	3 in	1.5 in	0.5 in	1.5 in
2		Part2-03	Part2-03	0.75 in	5 in	1.5 in	0.75 in	3 in
3		Part2-02	Part2-02	0.75 in	3 in	1.5 in	0.5 in	1.5 in

11. In the third row of the table, type-in new values in (2, 0.75, and 3) the Headsize, Headwidth, and Threadlength boxes. This creates the third version of the bolt.

		Member	Part Number	Diameter	Height	Headsize	Headwidth	Threadlenght
1		Part2-01	Part2-01	0.75 in	3 in	1.5 in	0.5 in	1.5 in
2		Part2-03	Part2-03	0.75 in	5 in	1.5 in	0.75 in	3 in
3		Part2-02	Part2-02	0.75 in	3 in	2 in	0.75 in	3 in

Now, you have to set the default version of the bolt.

12. Click the right mouse button on the third row of the table and select **Set As Default Row**.

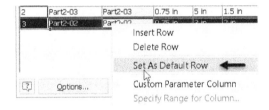

13. Click **OK** to close the dialog. Notice that the default version of the bolt changes.

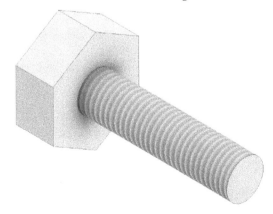

In the Bowser Bar, you will notice that the **Table** item is added.

14. Expand the **Table** item in the Browser Window to view the different variations of the iPart. Notice that the activated version of the iPart is designated by a check mark.

15. Double-click on any other version of the iPart to activate it.

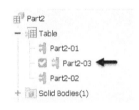

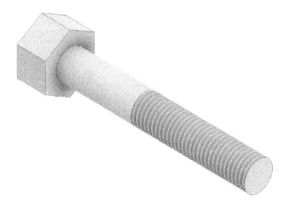

If you want to make changes to any version of the bolt, then click the right mouse button on it and select **Edit table**.

If you want to edit the table using a spreadsheet, then click the right mouse button on **Table** and select **Edit via Spreadsheet**. Click **OK** on the message box.

Now, modify the values in the spreadsheet and close it. A message pops up asking you to save the changes. Click **Save** to save the changes.

	A	B	C	D	E	F	G	H
1	Member	Part Num	Diameter	Height	Headsize	Headwidt	Threadler	Thread1
2	Part2-01	Part2-01	0.75 in	3 in	1.5 in	0.5 in	1.5 in	Compute
3	Part2-03	Part2-03	0.75 in	5	1.5 in	0.75 in	3	Compute
4	Part2-02	Part2-02	0.75 in	3 in	2	0.75 in	3	Compute

If you want to save anyone of the iPart versions as a separate part file, then click the right mouse button on it and select **Generate Files**.

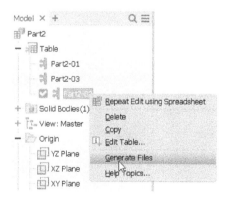

16. Save and close the file.

TUTORIAL 7

In this tutorial, you create a plastic casing.

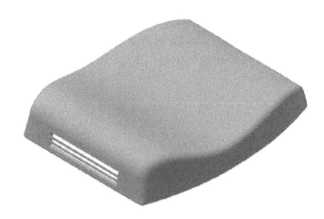

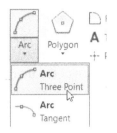

Creating the First Feature

1. Open a new Autodesk Inventor part file using the **Standard.ipt** template (See Chapter 2, Tutorial 3, Starting a New Part File section).
2. On the ribbon, click **3D Model > Sketch > Start 2D Sketch**.
3. Select the XZ Plane.
4. On the ribbon, click **Sketch > Create > Line**.
5. Click on the second quadrant, move the cursor horizontally toward the right, and then click to create a horizontal line, as shown.

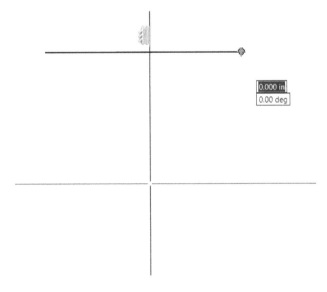

6. On the ribbon, click **Sketch > Create > Arc drop-down > Arc Three Point**.

7. Select the right endpoint of the horizontal line.
8. Move the cursor vertically downwards and click to define the second point.
9. Move the pointer towards the right and click on the horizontal axis line.

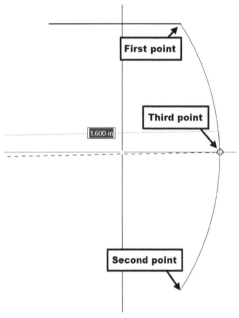

10. Likewise, create another three-point arc and horizontal line, as shown.

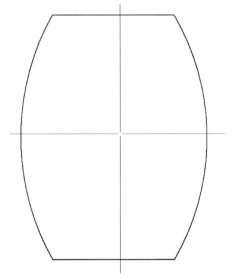

11. Create a vertical construction from the origin point.

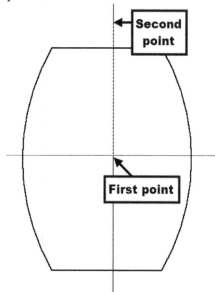

12. On the ribbon, click **Sketch > Constrain > Symmetric** .

13. Select the two arcs and the construction

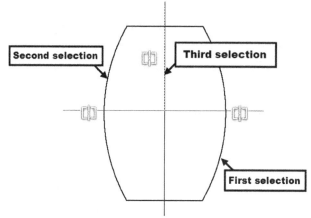

14. Likewise, create a horizontal construction line from the origin point, and then make the two horizontal lines symmetric about it.

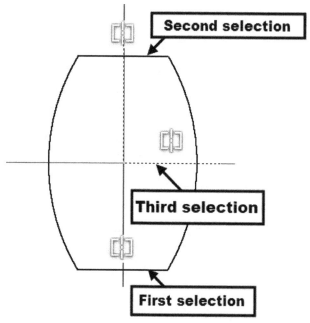

15. Add dimensions to the sketch. Also, add the Vertical constraint between the origin and the center point of the arc.

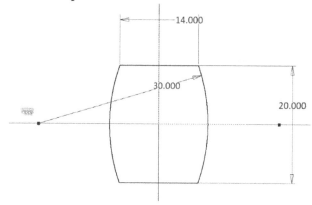

16. Click **Finish Sketch**.

17. Click the **3D Model > Create > Extrude** on the ribbon; the **Extrude Properties** panel dialog appears.

18. Set the **Distance A** to 3.15.

19. Expand the **Advanced Properties** section and set the Taper angle to -10.

20. Click the **OK** button.

Creating the Extruded surface

1. Click **3D Model > Sketch > Start 2D Sketch** on the ribbon, and then select the XY Plane.

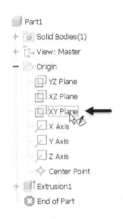

2. Click the **Slice Graphics** button at the bottom of the window or press **F7** on the keyboard.
3. Click **Sketch > Create > Line > Spline Interpolation** on the ribbon.
4. Create a spline, as shown in figure (See Chapter 5, Tutorial 4, Creating the First Section and Rails section).

5. Apply dimensions to the spline, as shown below.

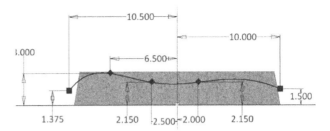

Adding dimension between two points of a spline locks the distance between the two points. Also, removing the dimension will revert the spline to its original shape.

6. Click **Finish Sketch**.

7. Click the **Extrude** button.

8. Select the sketch, if not already selected.
9. On the **Extrude Properties** panel, click the toggle button on the top right corner of the **Extrude Properties** panel.

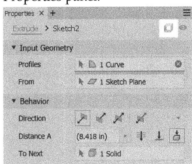

26. Select the **Symmetric** button under the **Behavior** section.
27. Extrude the sketch up to 17 in **Distance A**.

Replacing the top face of the model with the surface

1. On the **Surface** panel of the **3D Model** ribbon, click the **Replace Face** 🔳 button; the **Replace Face** dialog appears.

Now, you need to select the face to be replaced.

2. Select the top face of the model.

Next, you need to select the replacement face or surface.

3. Click the **New Faces** button on the dialog and select the extruded surface.

 *You can also use a solid face to replace an existing face.*

4. Click **OK** to replace the top face with a surface.

5. Hide the extruded surface by clicking the right mouse button on it in the Browser Window and un-checking **Visibility**.

Creating a Face fillet

1. Click the **Fillet** button on the **Modify** panel.
2. Click the **Face Fillet** button on the **Fillet** dialog.

3. Select the top surface as the first face and the

inclined front face as the second face.

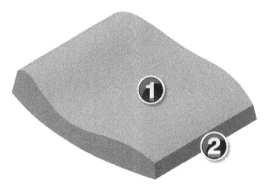

4. Set the **Radius** to 1.5 and click the **OK** button to create the face fillet.

5. Likewise, apply a face fillet of 1.5 radius between the top surface and the inclined back face of the model.

Creating a Variable Radius fillet

1. Click the **Fillet** button on the **Modify** panel.

2. Click the **Variable** tab on the **Fillet** dialog.

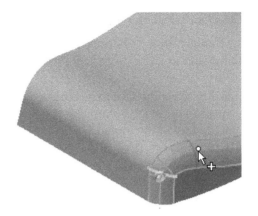

3. Select the curved edge on the model; the preview of the fillet appears.

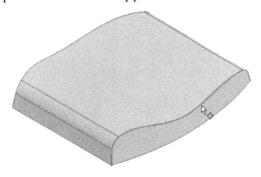

4. Select a point on the fillet, as shown in the figure.

6. Set the radii of the **Start**, **End**, **Point 1**, and **Point 2**, as shown below.

Point	Radius	Position
Start	.6	0.0
End	.6	1.0
Point 1	1	0.0000
Point 2	.8	0.9103

You can also specify the fillet continuity type. By default, the **Tangent Fillet** type is specified.

6. Select **Smooth (G2) Fillet** type from the **Edges** section.

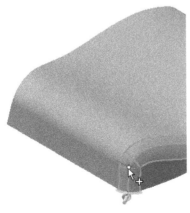

5. Select another point of the fillet, as shown in the figure.

7. Make sure that the **Smooth radius transition** option is checked.
8. Click **OK** to create the variable fillet.

Mirroring the fillet

9. Click the **Mirror** ⚠ button on the **Pattern** panel; the **Mirror** dialog appears.
10. Select the variable radius fillet from the model.
11. Click the **Mirror Plane** button on the dialog.
12. Select the **XY Plane** from the Browser window.
13. Click **OK** to mirror the fillet.

Shelling the Model

1. Click the **Shell** ⬚ button on the **Modify** panel; the **Shell** dialog appears.
2. Click the **Inside** 🔲 button on the dialog and set the **Thickness** to 0.2 in.
3. Click the **Remove Faces** icon, if not already selected.
4. Rotate the model and select the bottom face.

5. Click **OK**.

Creating the Boss Features

1. Click **3D Model > Sketch > Start 2D Sketch** on the ribbon and select the bottom face of the model.

2. Activate **Construction** ⟍ button on the **Format** panel.
3. On the ribbon, click **Sketch > Create > Rectangle drop-down > Rectangle Two Point Center**.

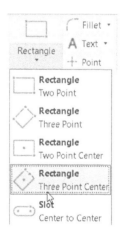

4. Select the origin point of the sketch.
5. Move the cursor outward and click to create the rectangle.
6. Apply dimensions to the rectangle.
7. Click the **Point** button on the **Create** panel.
8. Place four points at corners of the rectangle.

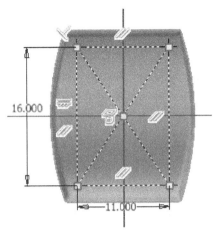

9.   Click **Finish Sketch**.

Now, you will create bosses by selecting the points created in the sketch.

10. On the ribbon, click the **Show Panels** ⊙ ˇ button located at the right side, and then select **Plastic Part** from the menu.

11. Click the **Boss** button on the **Plastic Part** panel; the **Boss** dialog appears.

12. Click the **Thread** button on the dialog.

13. Select the **From Sketch** option from the **Placement** group.

14. Select the points located on the corners of the rectangle, if not already selected; the bosses are placed at the selected points.

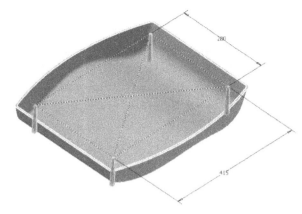

15. Click the **Thread** tab and specify the parameters, as shown below.

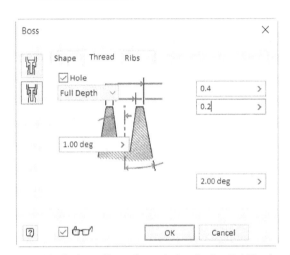

16. Click the **Ribs** tab and check the **Stiffening Ribs** option.

17. Set the rib parameters, as shown next.

132

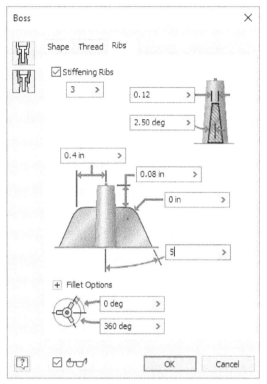

18. Expand the **Fillet options**.

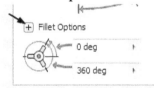

19. Specify the fillet options, as shown below.

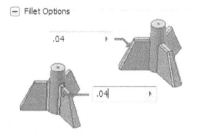

20. Click **OK** to create the bosses with ribs.

Creating the Lip feature

1. Click the **Lip** button on the **Plastic Part** panel of the ribbon; the **Lip** dialog appears.
2. Click the **Lip** button on the dialog.

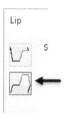

3. Select the outer edge of the bottom face.

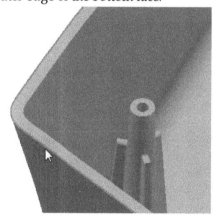

4. Click the **Guide Face** button on the dialog and select the bottom face of the model.

5. Click the **Lip** tab and set the parameters, as shown below.

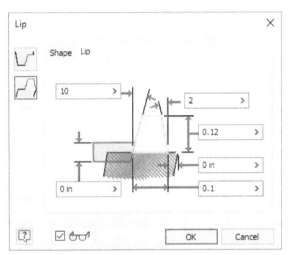

6. Click **OK** to create the lip.

Creating the Grill Feature

1. Click the **Home** button located at the top of the ViewCube.
2. Click the corner point of the ViewCube, as shown.

3. On the ribbon, click **3D Model > Sketch >> Start 2D Sketch**.
4. Select the inclined face, as shown.

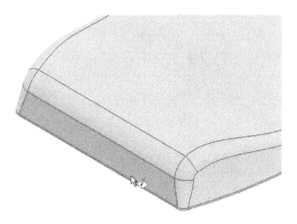

5. Create the sketch using the **Rectangle Two Point Center** and **Line** tools.

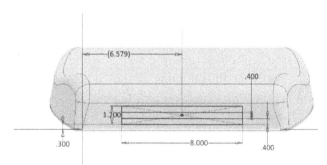

6. Click **Finish Sketch**.
7. Click the **Grill** button on the **Plastic Part** panel.

8. Select the rectangle as the boundary and set the **Boundary** parameters, as shown below.

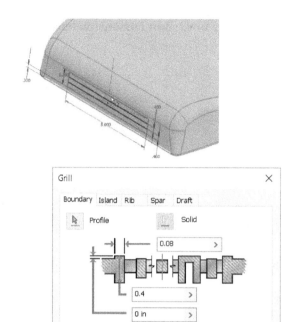

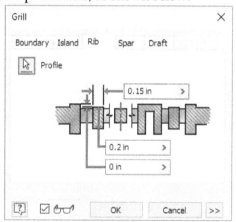

9. Click the **Rib** tab and select the horizontal lines.
10. Set the rib parameters, as shown below.

11. Click **OK** to create the grill.

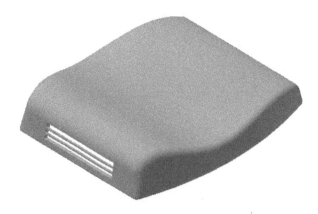

12. Save the model as Plastic Cover.ipt.

Creating Ruled Surface

1. Click **3D Model > Surface > Ruled Surface** on the ribbon and select the bottom edge of the model.

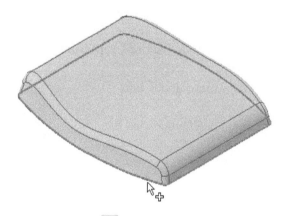

2. Click the **Normal** button on the **Ruled Surface** dialog.

   The preview of the ruled surface appears normal to the selected edge.

   You can click the **Alternate All Faces** button to change the direction of the ruled surface.

3. Type in 2 in the **Distance** box.
4. Click **OK** to create the ruled surface.

The ruled surface can be used as a parting split while creating a mold.

5. Close the part file without saving.

TUTORIAL 8 (The Extent Start option)

In this tutorial, you will learn the use of the **Extent Start** option in the **Hole** tool.

1. Download the Tutorial_9.ipt from the companion website: www.tutorialbook.info

2. Open the downloaded file.

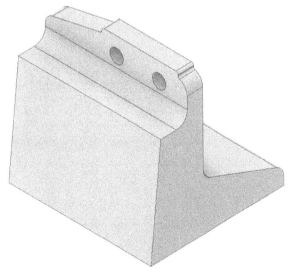

3. In the Browser Window, expand the **Hole** feature, and then right click on the **Sketch**.
4. Select **Edit Sketch** from the Shortcut Menu; the sketch is displayed.
5. Double click on the 0.19 dimension.

6. Type 0.12 in the **Edit Dimension** box, and then click the green check.

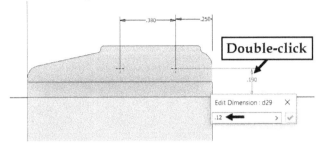

7. Click **Finish Sketch** on the ribbon.

Notice that the fillet overlaps with the holes.

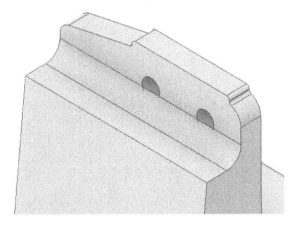

8. In the Browser Window, right click on the **Hole** feature, and then select **Edit Feature**.
9. On the **Properties** panel, expand the **Advanced Properties** section and check the **Extend Start** option.

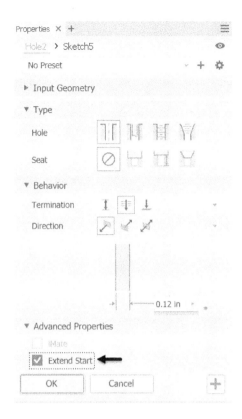

10. Click **OK** to close the dialog.

The **Extend Start** option removes the overlapping material.

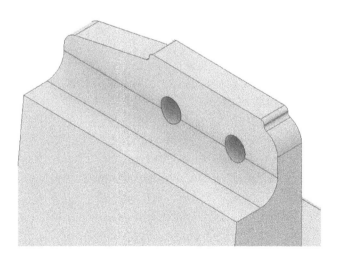

TUTORIAL 10 (Partial chamfer)

In this tutorial, you will learn to create a partial chamfer.

1. Start a new part file using the **Standard(in).ipt** template.

2. Create a 1 X 1 X 1 box using the Extrude tool, as shown.

3. On the ribbon, click **3D Model** tab > **Modify** panel > **Chamfer**.

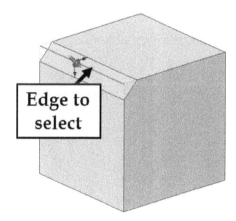

4. On the **Chamfer** dialog, click the **Partial** tab, and then click on the selected edge, as shown.

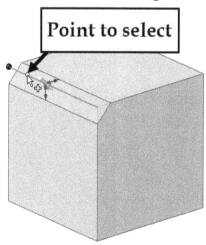

5. On the **Partial** tab, select **Set the Driven Dimension > To End**.
6. Change the **To Start** and **Chamfer** values to 0.25 and 0.5, respectively.
7. Click **OK**.

8. Save and close the part file.

TUTORIAL 11

In this tutorial, you construct a patterned cylindrical shell.

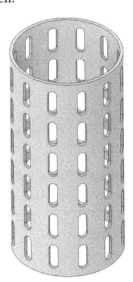

Constructing a cylindrical shell

1. Start a new part file.
2. Create a sketch on the XZ plane.

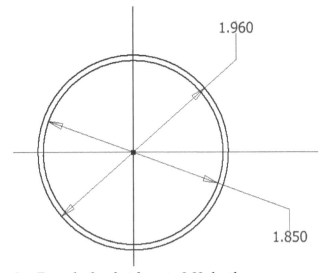

3. Extrude the sketch up to 3.93 depth.

Adding a Slot

1. Activate the **Start 2D Sketch** tool.
2. In the Browser Window, expand the **Origin** folder and select the XY Plane.
3. Click the **Slice Graphics** icon on the Status bar.

The model is sliced using the sketching plane.

4. On the ribbon, click **Sketch** tab > **Create** panel > **Rectangle** drop-down > **Slot Center to Center** 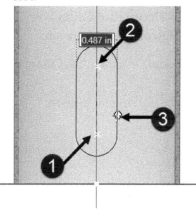.

5. Click to define the first point of the slot.

6. Move the pointer up and click to define the second point.

7. Move the pointer outward and click to create a slot.

8. Apply the Vertical constraint between the Center point of the slot and the origin point.

9. Add dimensions to the slot.

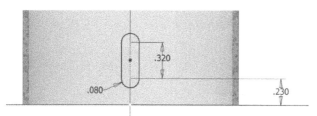

10. Click **Finish Sketch**.

11. On the ribbon, click **3D Model** tab > **Create**

panel > **Extrude**.

12. On the **Extrude Properties** panel, click the **Through All** icon under the **Behavior** section.

13. Make sure that the arrow on the preview points in the forward direction.

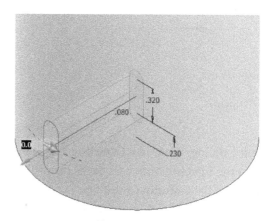

14. Click the **Cut** icon on the dialog.

15. Click **OK**.

Constructing the Rectangular pattern

1. On the ribbon, click **3D Model** > **Pattern** > **Rectangular Pattern** .

2. On the **Rectangular Pattern** dialog, click the **Features** button, and then select the slot.

3. On the **Rectangular Pattern** dialog, click the cursor button in the **Direction 1** section.

4. In the **Browser Window**, expand the **Origin** folder, and then select the Y Axis.

5. On the dialog, under the **Direction 1** section, select the **Spacing** option from the drop-down.
6. Type **6** in the **Column Count** box.
7. Type **0.629** in the **Colum Spacing** box.

8. Click **OK**.

Constructing the Circular pattern

1. On the ribbon, click **3D Model > Pattern > Circular Pattern** .
2. In the Browser Window, select the rectangular pattern.
3. On the **Circular Pattern** dialog, click the **Rotation Axis** box and select the cylindrical face of the model.
4. Type-in **12** in the **Occurrence Count** box.
5. Type **360** in the **Occurrence Angle** box.
6. Click the **Rotational** icon in the **Orientation** section.
7. Click **OK** to make the circular pattern.

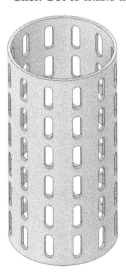

11. Save and close the model.

TUTORIAL 12

In this tutorial, you create the model shown in the figure.

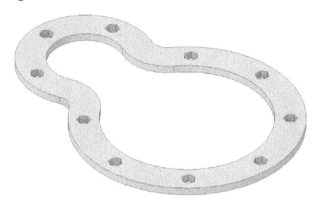

Constructing the first feature

1. Open a new part file.
2. On the ribbon, click **3D Model > Sketch > Start 2D Sketch**.
3. Click on the XZ plane.
4. Construct two circles and add dimensions to them.

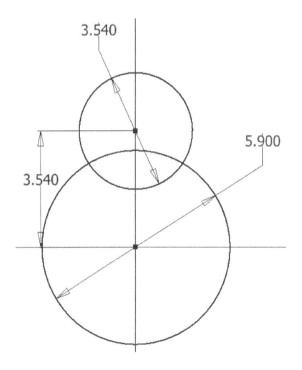

5. On the ribbon, click **Sketch > Modify > Trim** .
6. Click and drag the left mouse across the intersections of the two circles.

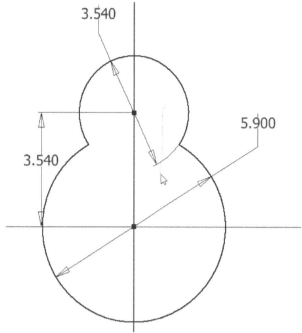

7. On the ribbon, click **Sketch > Create > Fillet** and set the **Radius** value to 0.39.
8. Make sure that the **equal to** button is pressed on the **2D Fillet** dialog.
9. Click on the two arcs; a fillet is created at one corner.
10. Again, click on the two arcs.
11. Right click and click **OK**.

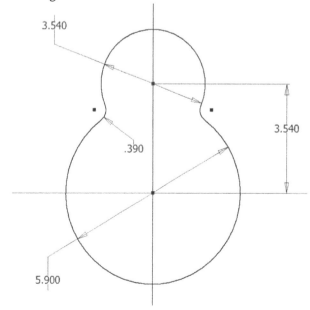

12. On the ribbon, click **Sketch** tab > **Modify** panel > **Offset**.

13. Click on the sketch, and then move the pointer inward.
14. Type 0.78 in the box attached to the offset element.
15. Press Enter.
16. Click **Finish Sketch**.
17. On the ribbon, click **3D Model > Create > Extrude**.
18. Click inside the region enclosed by the sketch and its offset.

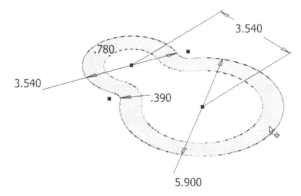

19. On the **Extrude** Properties panel, type 0.196 in the **Distance A** box.
20. Click **OK**.

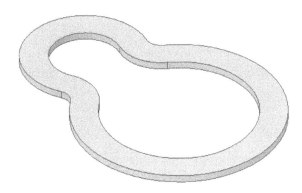

Constructing the Extruded cut

1. On the ribbon, click **3D Model > Sketch > Start 2D Sketch**.
2. Click on the top face of the model geometry.
3. On the ribbon, click **Sketch > Create > Rectangle** drop-down **> Polygon** .
4. On the dialog, type-in 6 in the **Number of Sides** box.
5. Click the **Inscribed** icon.

6. Click to define the center point of the polygon.

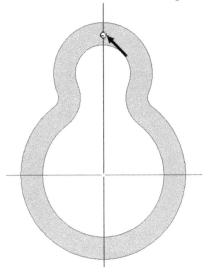

7. Move the pointer and click to construct the polygon.
8. Right click and select **OK**.
9. On the ribbon, click **Sketch > Constrain > Horizontal**.
10. Click on the line of the polygon, as shown.

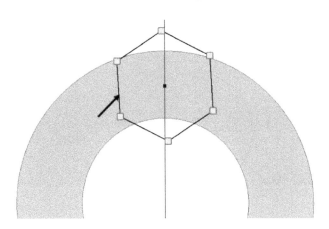

11. Select the center point of the polygon and the sketch origin.

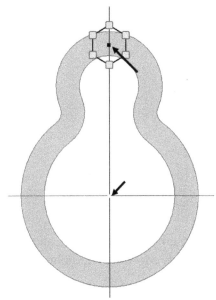

12. Activate the **Dimension** command, zoom into the polygon, and select the two lines, as shown.

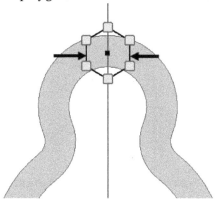

13. Place the dimension, type 0.31, and press Enter.
14. Select the center point of the polygon and the sketch origin.
15. Place the dimension, and type 4.92. Next, press Enter.

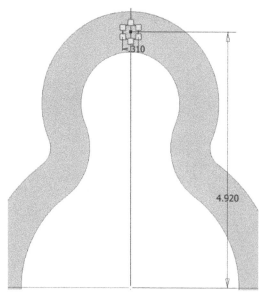

16. Click **Finish Sketch**.
17. On the ribbon, click **3D Model > Create > Extrude**.
18. Click on the region enclosed by the polygon.
19. Click the **Cut** icon on the **Extrude Properties** panel.
20. Click the **Through All** icon under the **Behavior** section.
21. Click **OK**.

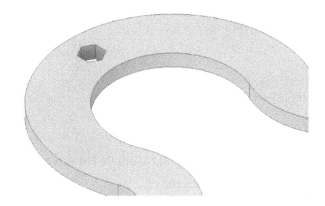

Making the Path Pattern

1. On the ribbon, click **3D Model > Sketch > Start 2D Sketch**.
2. Click on the top face of the model.
3. On the ribbon, click **Sketch > Create > Project Geometry** .
4. Right click on the outer edge of the model, and

then select **Select Tangencies**.

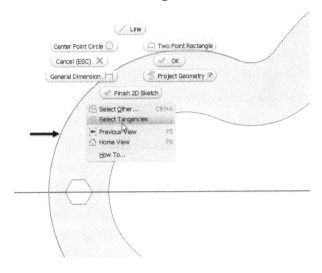

5. On the ribbon, click **Sketch > Create > Point**.
6. Specify the point, as shown.

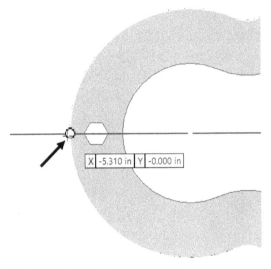

7. Right click and select **OK**.
8. Click **Finish Sketch**.
9. On the ribbon, click **3D Model > Pattern > Rectangular Pattern** .
10. Select the Extruded cut feature from the model.

11. Click the cursor button in the **Direction 1** section.
12. Place the pointer on the projected curve, and then click when all its edges are highlighted in red.

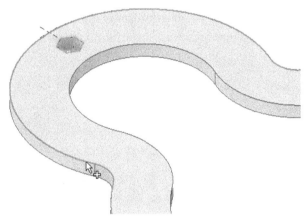

13. Click the expand button on the **Rectangular Pattern** dialog.
14. Click the **Start** icon in the **Direction 1** section.
15. Select the point created in the sketch to define the start point of the pattern.

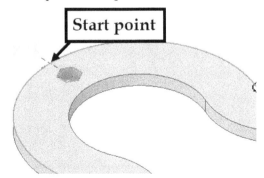

Start point

16. Type 10 in the **Column Count** box.
17. Select the **Curve Length** from the drop-down available in the **Direction 1** section.
18. Select **Direction 1** from the **Orientation** section.

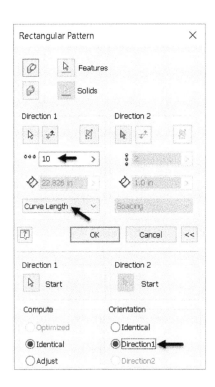

19. Click **OK** to create the pattern along a path.

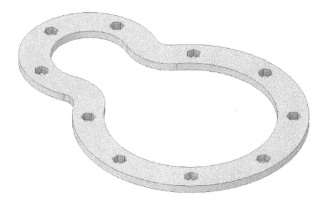

20. Save and close the file.

Chapter 7: Sheet Metal Modeling

This chapter will show you to:

- Create a face feature
- Create a Flange
- Create a Contour Flange
- Create a Corner Seam
- Create Punches
- Create a Bend Feature
- Create Corner Rounds
- Flat Pattern

TUTORIAL 1

In this tutorial, you create the sheet metal model shown in the figure.

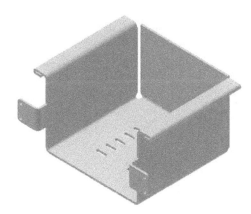

Starting a New Sheet metal File

1. To start a new sheet metal file, click **Get Started Launch > New** on the ribbon.
2. On the **Create New File** dialog, click the **Sheet Metal.ipt** icon, and then click **Create**.

Sheet
Metal.ipt

Setting the Parameters of the Sheet Metal part

1. To set the parameters, click **Sheet Metal > Setup > Sheet Metal Defaults** on the ribbon; the **Sheet Metal Defaults** dialog appears.

Sheet Metal
Defaults

Setup ▼

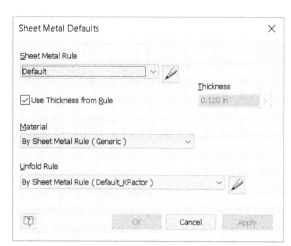

This dialog displays the default preferences of the sheet metal part such as sheet metal rule, thickness, material, and unfold rule. You can change these preferences as per your requirement.

2. To edit the sheet metal rule, click the **Edit Sheet Metal Rule** button on the dialog.

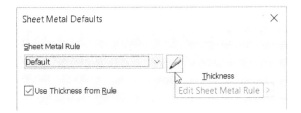

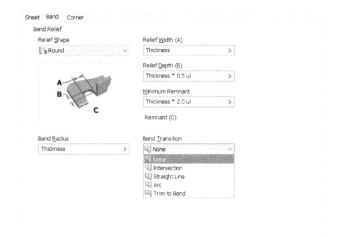

In the **Sheet** tab of the **Style and Standard Editor** dialog, you can set the sheet preferences such as sheet thickness, material, flat pattern bend angle representation, flat pattern punch representation, and gap size.

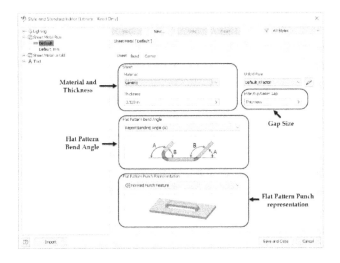

3. In the **Sheet** tab, set the **Thickness** to 0.12 and leave all the default settings.
4. Click the **Bend** tab.

In the **Bend** tab of this dialog, you can set the bend preferences such as bend radius, bend relief shape and size, and bend transition.

5. Set the **Relief Shape** to **Round**.
6. Click the **Corner** tab.

In the **Corner** tab, you can set the shape and size of the corner relief to be applied at the corners.

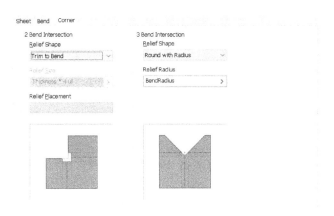

7. After setting the required preferences, click the **Save and Close** button.

The **Unfold Rule** option on the **Sheet Metal Defaults** dialog defines the folding/unfolding method of the sheet metal part. To modify or set a new Unfold Rule, click the **Edit Unfold Rule** button on the **Sheet Metal Defaults** dialog.

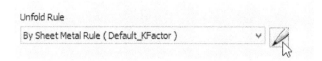

On the **Style and Standard Editor** dialog, select the required **Unfold Method**.

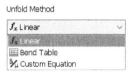

You can define the Unfold rule by selecting the **Linear** method (specifying the K factor), selecting a **Bend Table**, or entering a custom equation. Click **Save and Close** after setting the parameters.

8. Close the **Sheet Metal Defaults** dialog.

Creating the Base Feature

1. Create the sketch on the XZ Plane, as shown in figure (Use the **Rectangle Two Point Center** tool).

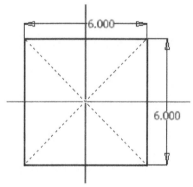

2. Click **Finish Sketch**.
3. To create the base component, click **Sheet Metal > Create > Face** on the ribbon; the **Face** dialog appears.

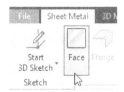

4. Click **OK** to create the tab feature.

Creating the flange

1. To create the flange, click **Sheet Metal > Create > Flange** on the ribbon; the **Flange** dialog appears.

2. Select the edge on the top face, as shown.

3. Set the **Distance** to 4.

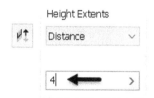

4. Click on the **Bend from the intersection of the two outer faces** icon in the **Height Datum** section. This measures the flange height from the outer face.

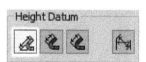

5. Under the **Bend Position** section, click the **Inside of Bend extents** icon.

6. Click **OK** to create the flange.

Creating the Contour Flange

1. Draw a sketch on the front face of the flange, as shown in the figure.

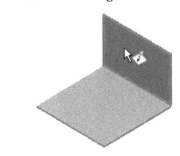

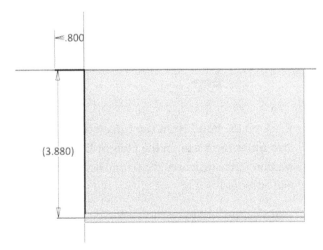

2. Click **Finish Sketch**.
3. To create the contour flange, click **Sheet Metal > Create > Contour Flange** on the ribbon; the

Contour Flange dialog appears.

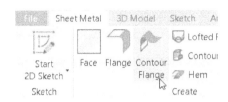

4. Select the sketch from the model.
5. Select the edge on the left side of the top face; the contour flange preview appears.

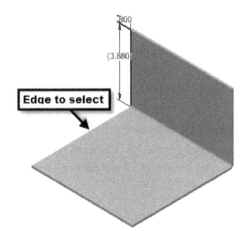

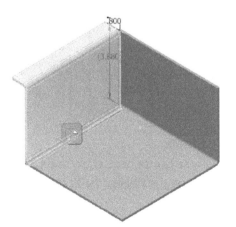

6. Select **Edge** from the **Type** drop-down.

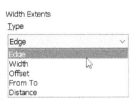

7. Click **OK** to create the contour flange.

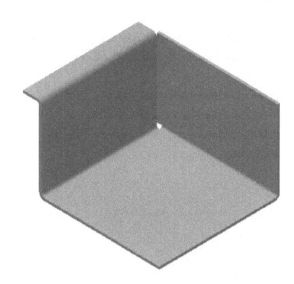

Creating the Corner Seam

1. To create the corner seam, click **Sheet Metal> Modify > Corner Seam** on the ribbon; the **Corner Seam** dialog appears.

2. Rotate the model.
3. Select the two edges forming the corner.

4. Set the parameters in the **Shape** tab of the dialog, as shown.

5. Click the **Bend** tab and make sure that the **Default** option is selected in the **Bend Transition** drop-down.
6. Click the **Corner** tab and set the **Relief Shape** to **Round**.

You can also apply other types of relief using the options in the **Relief Shape** drop-down.

7. Click **OK**.

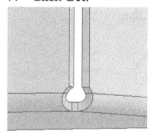

Creating a Sheet Metal Punch iFeature

1. Open a new sheet metal file using the **Sheet Metal.ipt** template.

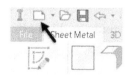

2. Create a sheet metal face of the dimensions 4x4.

3. Click **Manage > Parameters > Parameters** f_x on the ribbon; the **Parameters** dialog appears.

4. Select the **User Parameters** row and click the **Add Numeric** button on the dialog. This adds a new row.

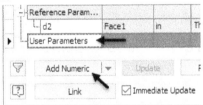

5. Enter **Diameter** in the new row.

6. Set **Unit Type** to **in** and type-in 0.04 in the **Equation** box.

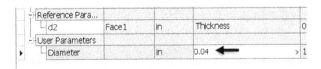

7. Likewise, create a parameter named **Length** and specify its values, as shown below.

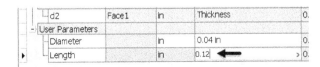

8. Click **Done**.

9. Click **Sheet Metal > Sketch > Start 2D Sketch** on the ribbon.

10. Select the top face of the base feature.

11. On the ribbon, click **Sketch > Create > Rectangle drop-down > Slot Center to Center**.

12. Click anywhere to define the first center point of the slot.

13. Move the cursor horizontally and click to define the second center point of the slot.

14. Move the cursor outward and click to define the slot radius.

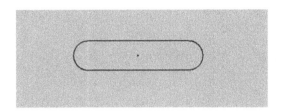

15. Click **Dimension** on the **Constrain** panel and select the round end of the slot.

16. Click to display the **Edit Dimension** box.

17. Click the arrow button on the box and select **List Parameters** from the shortcut menu; the **Parameters** list appears.

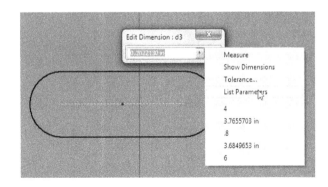

18. Select **Diameter** from the list and click the green check on the **Edit Dimension** box.

19. Likewise, dimension the horizontal line of the slot and set the parameter to **Length**.

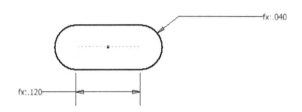

20. Click the **Point** button on the **Create** panel and place it at the center of the slot.
21. Delete any projected edges (yellow lines) from the sketch.
22. Click **Finish Sketch**.
23. Click **Sheet Metal > Modify > Cut** on the ribbon; the **Cut** dialog appears.

24. Accept the default values and click **OK** to create the cut feature.

25. Click **Manage > Author > Extract iFeature** on the ribbon; the **Extract iFeature** dialog appears.

26. On the dialog, select **Type > Sheet Metal Punch iFeature**.
27. Select the cut feature from the model geometry or from the Browser window. The parameters of the cut feature appear in the **Extract iFeature** dialog.

Next, you must set the **Size Parameters** of the iFeature.

28. Set the **Limit** of the **Diameter** value to **Range**. The **Specify Range** dialog appears.

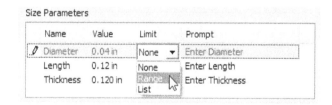

29. Set the values in the **Specify Range: Diameter** dialog, as shown below and click **OK**.

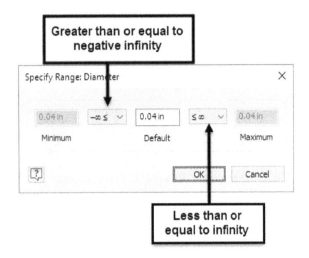

Next, you need to select the center point of the slot. This point will be used while placing the slot.

36. Click the **Select Sketch** button on the **Extract iFeature** dialog.

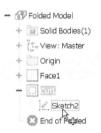

30. Set the **Limit** of the **Length** value to **List**. The **List Values: Length** dialog appears.
31. Click on **Click here to add value** and enter 0.2 as value.
32. Likewise, type-in values in the **List Values: Length** dialog, as shown below.

37. Select the sketch of the cut feature from the Browser window.

38. Click **Save** on the dialog; the **Save As** dialog appears.
39. Browse to the **Punches** folder and enter **Custom slot** in the **File name** box.

33. Click **OK**.
34. Set the **Limit** of the **Thickness** value to **Range**. The **Specify Range: Thickness** dialog appears.
35. Set the values in the **Specify Range: Thickness** dialog, as shown below. Next, click **OK**.

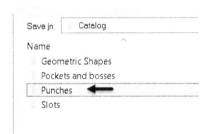

40. Click **Save** and **Yes**.
41. Click **File Menu > Save**.
42. Save the sheet metal part file as Custom slot.
43. Switch to the sheet metal file of the current

tutorial.

Creating a Punched feature

1. Start a sketch on the top face of the base sheet.
2. On the ribbon, click **Sketch > Create > Point**.
3. Place a point and add dimensions to it, as shown below.

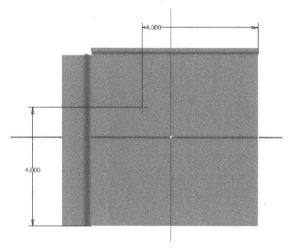

4. Click **Finish Sketch**.
5. To create the punch, click **Sheet Metal > Modify > Punch Tool** on the ribbon; the **PunchTool Directory** dialog appears.

6. Select Custom slot.ide from the dialog and click **Open**; the **PunchTool** dialog appears.

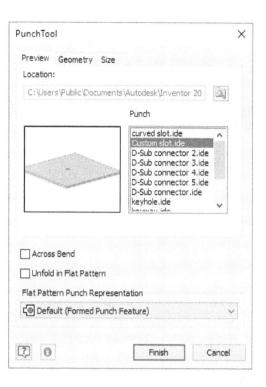

7. Click the **Size** tab on the **PunchTool** dialog.

8. Set **Length** to 0.45 and **Diameter** to 0.1.

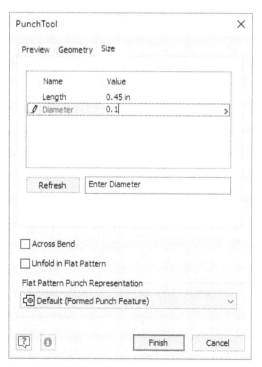

9. Click **Refresh** to preview the slot.
10. Click **Finish** to create the slot.

Note: If the slot is not oriented as shown in the figure, then click the **Geometry** tab on the **PunchTool** dialog and type in **90** in the **Angle** box.

Creating the Rectangular Pattern

1. Click **Sheet Metal > Pattern > Rectangular Pattern** on the ribbon. The **Rectangular Pattern** dialog appears.

2. Select the slot feature.

*You can also select multiple solid bodies from the graphics window using the **Pattern Solids** option.*

3. Click the **Direction 1** button on the dialog.

4. Select the edge of the base feature, as shown below.

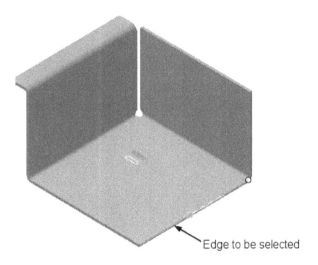

Edge to be selected

5. Select **Spacing** from the drop-down located in the **Direction 1** group.
6. Specify **Column Count** as 5.
7. Specify **Column Span** as 0.6.

8. Click the **Direction 2** button on the dialog.

9. Select the edge of the base feature, as shown below.

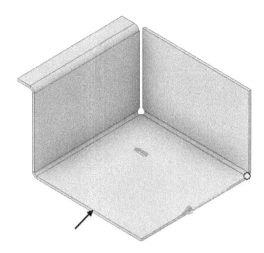

10. Click the **Flip** button in the **Direction 2** section to make sure the arrow is pointed toward the right.
11. Select **Spacing** from the drop-down located in the **Direction 2** group.
12. Specify **Column Count** as **2**.
13. Specify **Column Span** as 2.

14. Click **OK** to create the pattern.

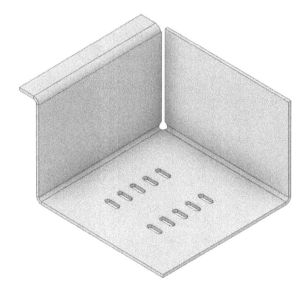

Creating the Bend Feature

1. Create a plane parallel to the front face of the flange feature. The offset distance is 6.3.

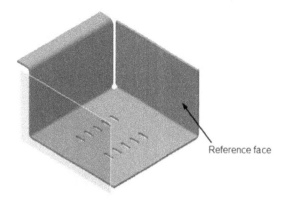

Reference face

2. Create a sketch on the new work plane.

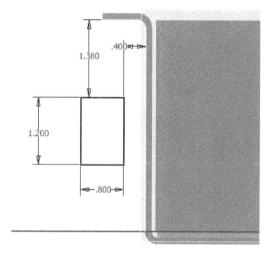

3. Click **Finish Sketch**.
4. Click **Sheet Metal > Create > Face** on the ribbon and create a face feature.

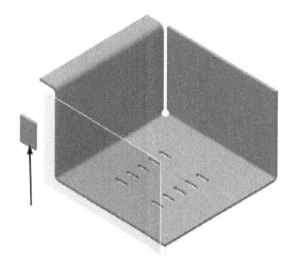

5. Click **Sheet Metal > Create > Bend** on the ribbon. The **Bend** dialog appears.

6. Select the edges from the model, as shown below.

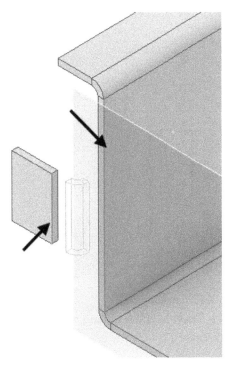

7. Make sure the **Bend Extension** is set to perpendicular.

8. Click **OK** to create the bend feature.

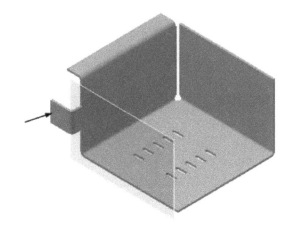

9. Hide the work plane (Right-click on it and uncheck **Visibility**).

Applying a corner round

1. To apply a corner round, click **Sheet Metal > Modify > Corner Round** on the ribbon; the **Corner Round** dialog appears.

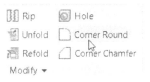

2. Set the **Radius** value to 0.2.
3. Set the **Select Mode** to **Feature**.

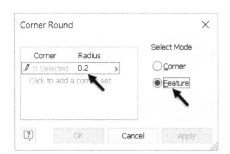

4. Select the face feature from the model.

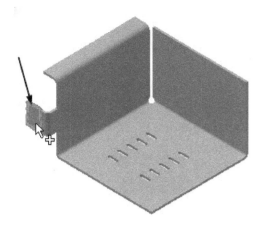

5. Click **OK** to apply the rounds.

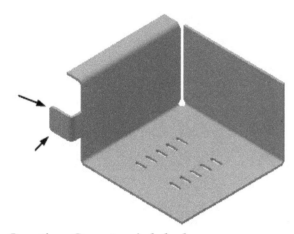

Creating Countersink holes

1. Click **Sheet Metal > Modify > Hole** on the ribbon; the **Properties panel** appears.

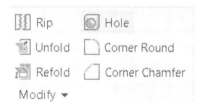

2. Set the **Placement** method to **Concentric**.
3. Set the hole type to **Countersink**.
4. Set the other parameters on the Hole Properties panel, as shown below.

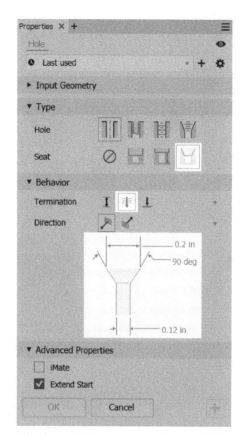

5. Click on the face of the flange, as shown below.

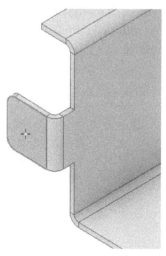

6. Select the corner round as the concentric reference.

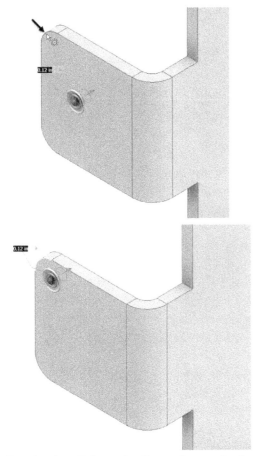

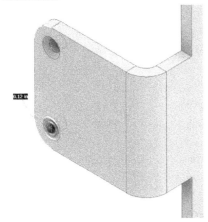

7. Again, click on the flange face and select the other rounded corner as the concentric reference.

8. Click **OK** to create the countersink.

Creating Hem features

1. To create the hem feature, click **Sheet Metal > Create > Hem** on the ribbon; the **Hem** dialog appears.

2. Set the **Type** to **Single**.

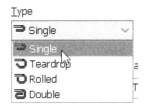

3. Select the edge of the contour flange, as shown below.

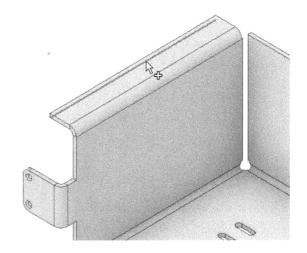

4. Leave the default settings of the dialog and click **OK** to create the hem.

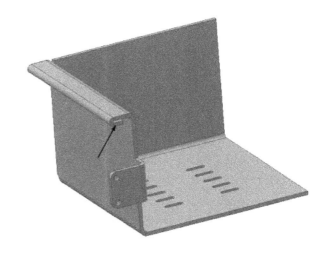

Mirroring the Features

1. Click **Mirror** ⬡ on the **Pattern** panel; the **Mirror** dialog appears.
2. Click **>>** at the bottom of the dialog and make sure the **Creation Method** is set to **Identical**.

3. Select the features from the Browser window, as shown below.

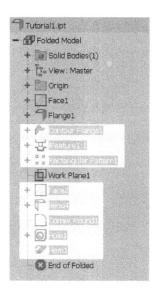

4. Click the **Origin YZ Plane** ⬡ button on the dialog
5. Click **OK** to mirror the feature.
6. Create a corner seam between the mirrored counter flange and flange.

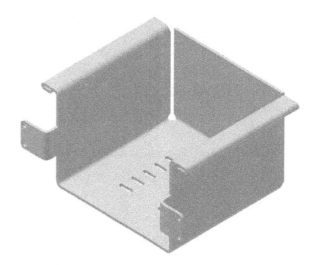

Creating the Flat Pattern

1. To create a flat pattern, click **Sheet Metal > Flat Pattern > Create Flat Pattern** on the ribbon.

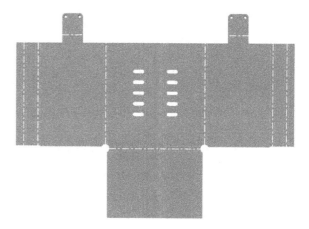

You can set the order in which the bends will be annotated.

2. Click the **Bend Order Annotation** button on the **Manage** panel of the **Flat Pattern** tab. The order in which the bends will be annotated is displayed.

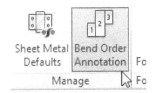

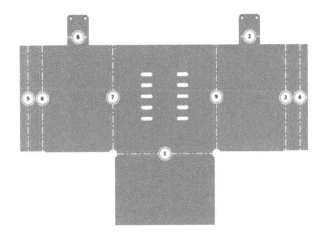

3. To change the order of the bend annotation, click on the balloon displayed on the bend. The **Bend Order Edit** dialog appears.
4. Select the **Bend Number** checkbox and enter a new number in the data field.

5. Click **OK** to change the order.
6. To switch back to the folded view of the model, click **Go to Folded Part** on the **Folded Part** panel.

7. Save the sheet metal part.

Creating 2D Drawing of the sheet metal part

1. On the **Quick Access toolbar**, click the **New** button.
2. On the **Create New File** dialog, double-click on **Standard.idw**.
3. Activate the **Base View** tool.
4. Click the Home icon on the ViewCube.
5. Leave the default settings on the **Drawing View** dialog and click **OK**.
6. Click and drag the drawing view to top right corner of the drawing sheet.

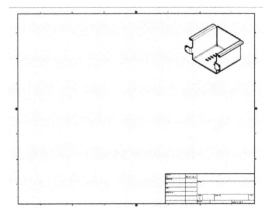

7. Likewise, create the front, and top views of the sheet metal part.

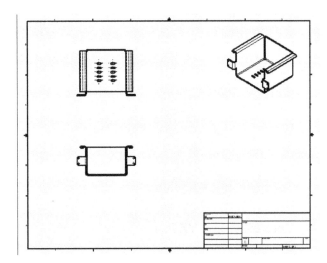

8. Activate the **Base View** tool and select **Sheet Metal View > Flat Pattern** on the **Drawing View** dialog.

9. Place the flat pattern view below the Isometric view.

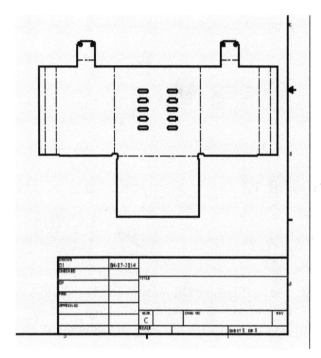

10. To add bend notes to the flat pattern, click **Annotate > Feature Notes > Bend** on the ribbon.

11. Click the horizontal bend line on the flat pattern to add the bend note.

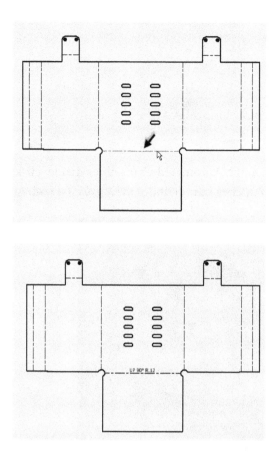

12. Likewise, select other bend lines on the flat pattern. You can also drag a selection box to select all the bend lines from the flat pattern view.

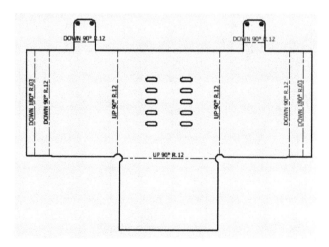

13. To add centerlines to the flat pattern view, click the right mouse button on it, and select **Automated Centerlines**.

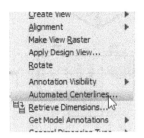

14. On the **Automated Centerlines** dialog, click the **Punches** button under the **Apply To** section.

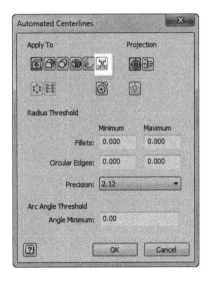

15. Click **OK** to add centerlines to the flat pattern view.

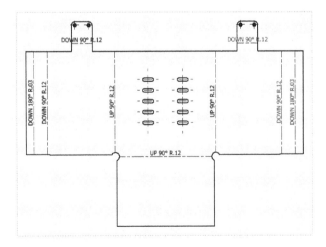

16. Likewise, add centerlines to other views on the drawing sheet.
17. To add a punch note, click **Annotate > Feature Notes > Punch** on the ribbon.

18. Zoom into the flat pattern view and click on the arc of the slot.
19. Move the pointer and click to create an annotation.

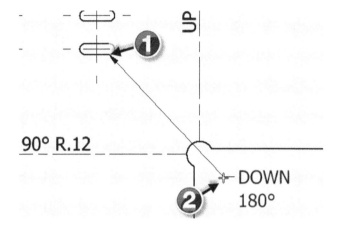

20. Use the **Retrieve Dimension** and **Dimension** tools to add dimensions to drawing.
21. Save and close the drawing and sheet metal part.

TUTORIAL 2

In this tutorial, you will convert an imported solid body into a sheet metal part.

1. Download the Tutorial_2. STEP file from the companion website.
2. On the Autodesk Inventor window, click the **Open** icon on the Quick Access Toolbar.
3. Browse to the location of the downloaded file, and then double-click on it.
4. On the **Import** dialog, select **Import Type > Convert Model**.
5. Click **OK** on the **Import** dialog.
6. On the ribbon, click the **3D Model** tab > **Convert > Convert to Sheet Metal** .
7. Click on the flat sheet metal face, as shown; the fixed face is defined.

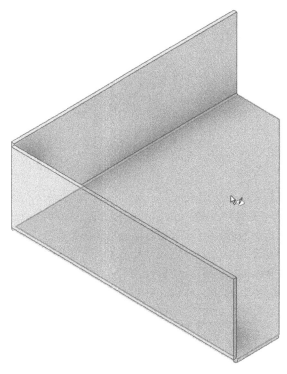

8. Set **Thickness** to 2
9. Click **OK**.
10. On the ribbon, click **Sheet Metal** tab > **Flat Pattern** panel > **Create Flat Pattern** .

TUTORIAL 3

In this tutorial, you will create a multibody sheet metal part.

1. Start a new Inventor part file.
2. Create a new face feature, as shown. The sheet metal thickness is 0.0787.

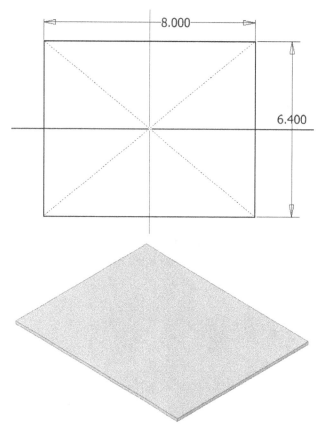

3. Create edge flanges of 1.575 inches length, as shown.

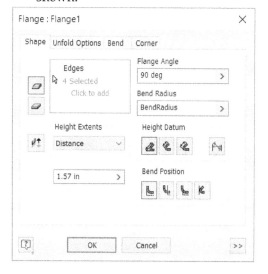

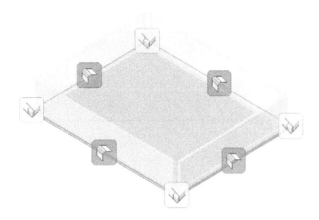

4. Activate the **Start 2D Sketch** tool, and then click on the outer face of the edge flange.

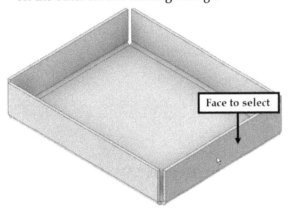

5. Create a rectangle, as shown.

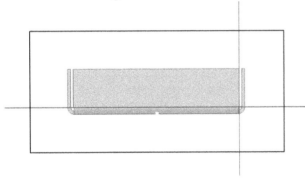

6. Press and hold the Ctrl key, and then select the bottom edge and the bottom horizontal line of the rectangle.

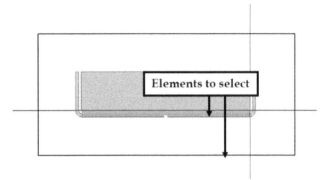

7. Click the **Collinear** icon the **Constrain** panel of the **3D Model** tab of the ribbon; the line is made collinear with the horizontal edge of the bend.
8. Likewise, make the vertical line collinear with the side edges, as shown.

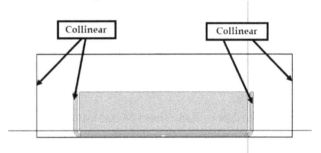

9. Add dimension to the vertical line, as shown.

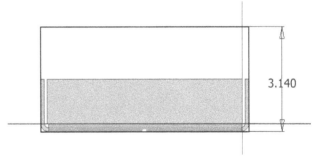

10. Click **Finish Sketch** on the ribbon.
11. Activate the **Face** command.
12. On the **Face** dialog, click the **New Solid** icon.
13. Make sure that the arrow on the face feature points outwards. Click the **Flip side** icon, if it points inwards.
14. Click **OK**; the face feature is created as a separate body.

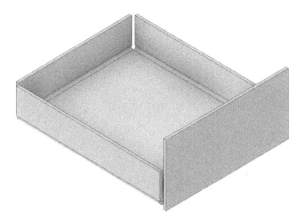

15. Expand the **Solid Bodies** folder; notice the two bodies.

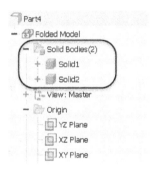

16. Click the top left corner of the ViewCube, as shown.

17. Activate the **Start 2D Sketch** tool.
18. Click on the flange face.

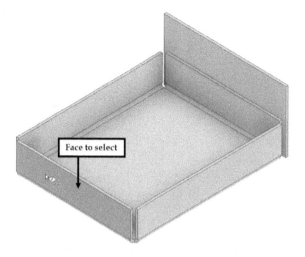

19. Activate the **Two Point Rectangle** tool.
20. Click on the first and second corners, as shown.

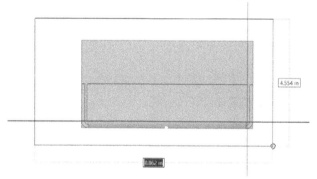

21. On the ribbon, click **Sheet Metal** tab > **Constrain** panel > **Coincident**.
22. Make the points coincident, as shown.

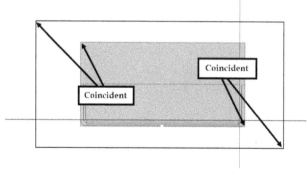

23. Click **Finish Sketch** on the ribbon.
24. Activate the Face command.
25. On the **Face** dialog, click the **New Solid** icon.
26. Click **OK**.

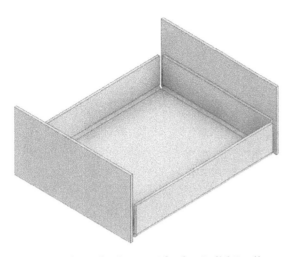

Notice three bodies inside the **Solid Bodies** folder in the **Model** tab of the Browser Window.

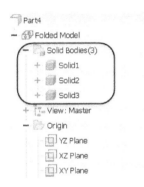

27. Right click and select **Home View**.
28. On the ribbon, click **Sheet Metal** tab > **Create** panel > **Bend** .
29. Click on the edges of the second and third sheet metal bodies, as shown.

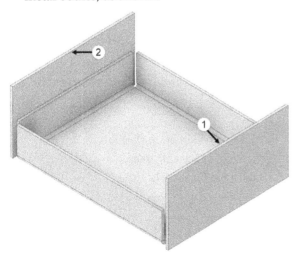

30. On the **Bend** dialog, select **Double Bend > 90**

Degree.
31. Click **OK** on the **Bend** dialog.

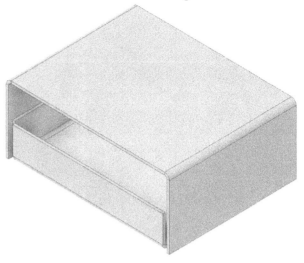

32. On the ribbon, click **Sheet Metal > Create > Flange** .
33. Click on the edge of the previously created edge flange, as shown.

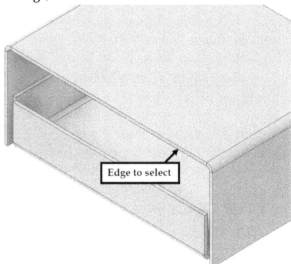

34. On the **Flange** dialog, click the **Shape** tab.
35. Select **Height Extents > To**.
36. Zoom to the bottom portion of the sheet metal part, and then select the vertex of the flange, as shown.

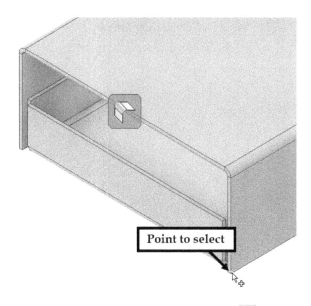

Point to select

37. Click the **Outside of base face extents** icon in the **Bend Position** group.
38. Click **OK**.
39. Likewise, create a flange on the opposite side.

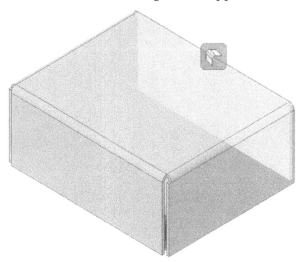

40. Save and close the sheet metal file.

TUTORIAL 4

In this tutorial, you will create a **Contour Roll** feature. The **Contour Roll** tool allows you to create a sheet metal feature by sweeping a profile about an axis.

1. Open a new Autodesk Inventor sheet metal file using the Sheet metal.ipt template.
2. On the ribbon, click **Sheet Metal** tab > **Start**

2D Sketch.
3. Click on XY Plane.
4. Draw the sketch profile, as shown.

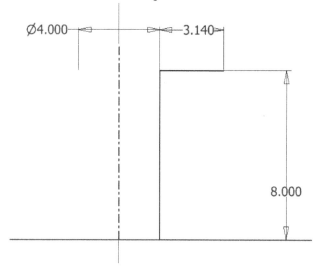

5. Click **Finish Sketch** on the ribbon.
6. On the ribbon, click **Sheet Metal > Create > Contour Roll** ; the profile and the axis are selected automatically.

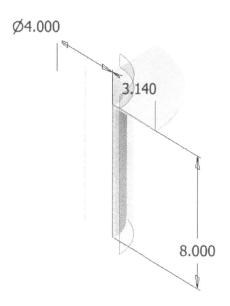

7. Specify the **Offset Direction** of the sheet metal (select **Both Sides** in this case).
8. Type **355** in the **Rolled Angle** box.
9. Specify the **Unroll Method**. There are three options **Centroid Cylinder**, **Custom Cylinder**, **Developed Length**, and **Neutral Radius** (select **Centroid Cylinder** in this

case).

10. Click **OK** to create the swept flange.

11. Save and close the part file.

Chapter 8: Top-Down Assembly and Joints

In this chapter, you will learn to

- Create a top-down assembly
- Insert Fasteners using Design Accelerator
- Export to 3D PDF
- Create assembly joints

TUTORIAL 1

In this tutorial, you will create the model shown in the figure. You use a top-down assembly approach to create this model.

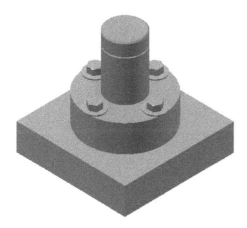

Creating a New Assembly File

1. To create a new assembly, click **New Assembly** on the Home screen.

Creating a component in the Assembly

In a top-down assembly approach, you create components of an assembly directly in the assembly by using the **Create** tool.

1. Click **Create** on the **Component** panel of the **Assembly** tab. The **Create In-Place Component** dialog appears.

2. Enter **Base** in the **New Component Name** field.

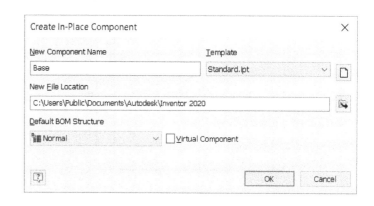

3. In the **Create In-Place Component** dialog, set the **New File Location** to the current project folder.

4. Click the **Browse to New File Location** icon.

5. On the **Save As** dialog, click the **Create New Folder** icon.

6. Type **C08_Tut_01** as the name of the folder.

7. Double-click on the new folder and click **Save**.

8. Click **OK** on the **Create In-Place Component** dialog.

9. Expand the **Origin** folder in the **Browser**

window and select the **XZ Plane**. The **3D Model** tab is activated in the ribbon.

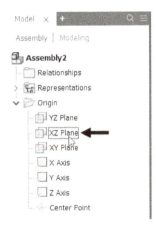

10. Click **Sketch > Start 2D Sketch** on the ribbon.
11. Select the **XZ Plane**.
12. Create a sketch, as shown below.

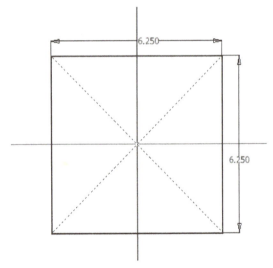

13. Click **Finish Sketch**.
14. Click **3D Model > Create > Extrude** on the ribbon and extrude the sketch up to 1.5 in.

15. Start a sketch on the top face and draw a circle of **2** in diameter.

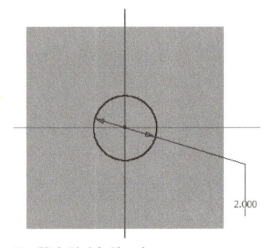

16. Click **Finish Sketch**
17. Extrude the sketch up to 3.75 inches distance.
18. Create a counterbore hole on the second feature (See Chapter 5, Tutorial 1, Create a Counterbore Hole section). The following figure shows the dimensions of the counterbore hole.

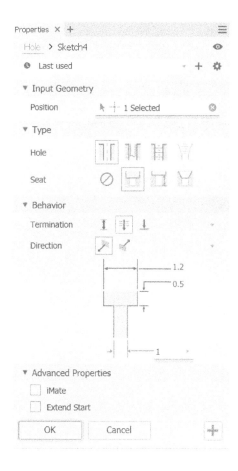

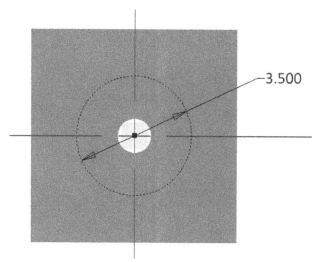

21. On the ribbon, click **Sketch > Create > Point**.
22. Place a point on the circle, as shown.

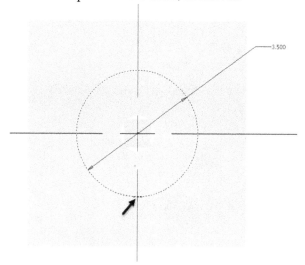

23. Click **Finish Sketch**.
24. On the ribbon, click **3D Model > Modify > Hole**.
25. On the **Properties** panel, specify the settings, as shown.

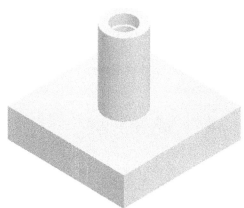

19. Start a new sketch on the top face of the first feature.
20. Create a 3.5 diameter circle with the **Construction** button active.

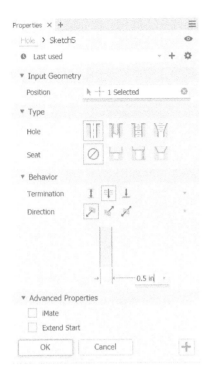

26. Make sure that the sketch point is selected.

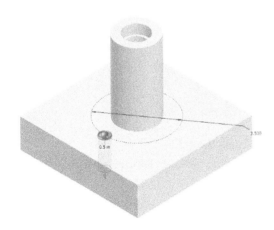

27. Click **OK** to create the hole.
28. Create a circular pattern of the hole (See Chapter 5, Tutorial 1, Create a Circular Pattern).

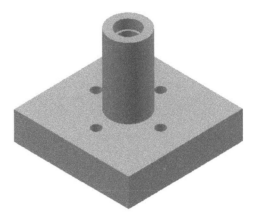

29. Click the **Return** button on the ribbon.

Creating the Second Component of the Assembly

1. Click **Assemble > Component > Create** on the ribbon; the **Create In-Place Component** dialog appears.
2. Enter **Spacer** in the **New Component name** field.
3. Check **Constrain sketch plane to selected face or plane** option.
4. Click **OK**.
5. Select the top face of the Base.

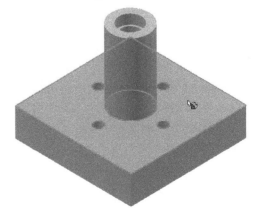

6. Click **Sketch > Start 2D Sketch** on the ribbon.
7. Select the top face of the Base.

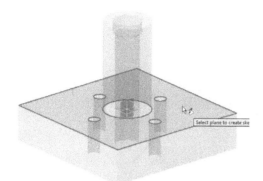

8. On the ribbon, click **Sketch > Create > Project Geometry** and select the circular edges of the Base.

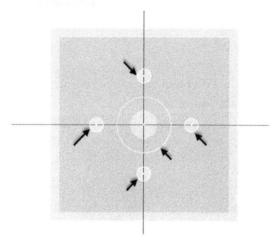

9. Draw a circle of 4.5 in diameter.

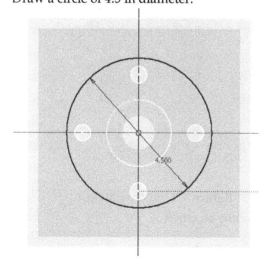

10. Click **Finish Sketch**.
11. Extrude the sketch up to 1.5 in.

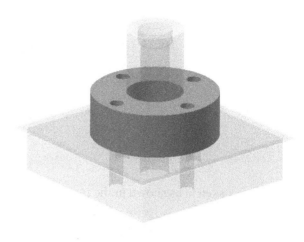

12. Click **Return** on the ribbon.

Creating the third Component of the Assembly

1. Click **Assemble > Component > Create** on the ribbon; the **Create In-Place Component** dialog appears.
2. Enter **Shoulder Screw** in the **New Component name** field.
3. Check **Constrain sketch plane to selected face or plane** option.
4. Click **OK**.
5. Click on the top face of the Base.

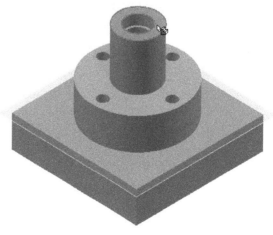

6. Start a sketch on the YZ Plane.

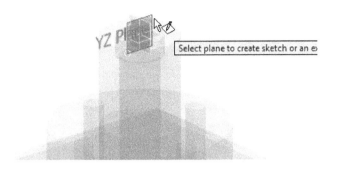

7. Draw a sketch, as shown in the figure.

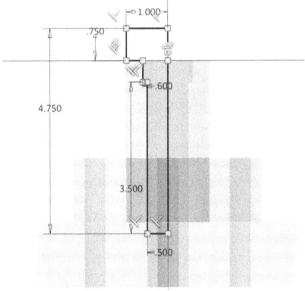

8. Click **Finish Sketch**.
9. Activate the **Revolve** tool and revolve the sketch.

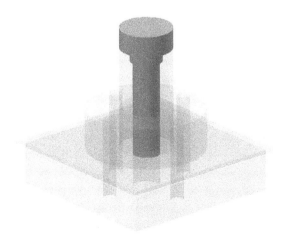

10. Activate the **Chamfer** tool and chamfer the

edges, as shown in the figure.

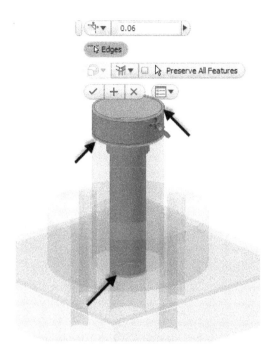

11. Activate the **Fillet** tool and round the edges, as shown in the figure.

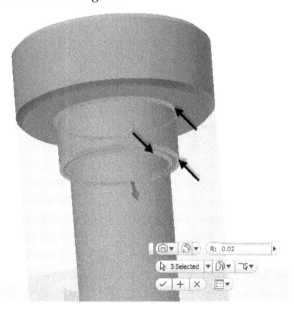

12. Click **Return** on the ribbon.
13. Save the assembly.

Adding Bolt Connections to the assembly

1. On the ribbon, click **Design > Fasten > Bolt Connection**.

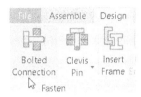

2. On the **Bolted Connection Component Generator** dialog, under the **Design** tab, select **Type > Through All**.

3. Select **Placement > Concentric**.

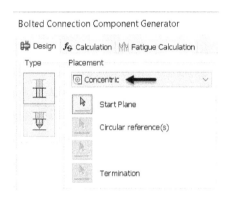

4. Select the top face of the Spacer.

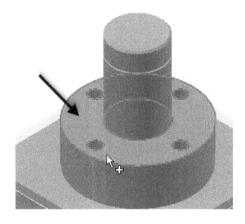

5. Click on the hole to define the circular reference.

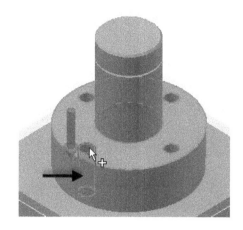

6. Rotate the model and click on the bottom face of the base. This defines the termination.

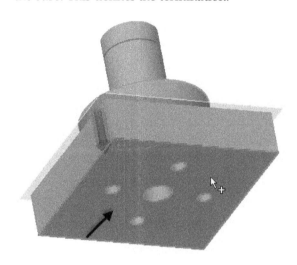

7. On the dialog, set the **Thread** type to **ANSI Unified Screw Threads**.

8. Make sure that the **Diameter** is set to **0.5** in.

9. On the dialog, click **Click to add a fastener**.

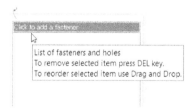

10. On the pop-up dialog, set the **Standard** to **ANSI** and **Category** to **Hex Head Bolt**.

11. Select **Hex Bolt-Inch**. This adds a hex bolt to the list.

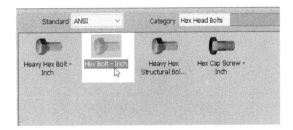

12. On the list, click **Click to add a fastener** below the Hex Bolt.

13. On the pop-up dialog, scroll down and select **Plain Washer (Inch)**.

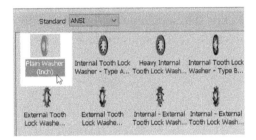

14. Click **Click to add a fastener** at the bottom of the list.

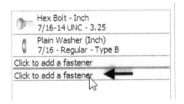

15. On the pop-up dialog, scroll down and select **Plain Washer (Inch)**.

16. Click **Click to add a fastener** at the bottom of the list.

17. On the pop-up dialog, set the **Category** to **Nuts** and select **Hex Nut –Inch**.

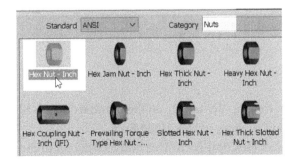

18. Click **OK** twice to add a bolt connection

subassembly.

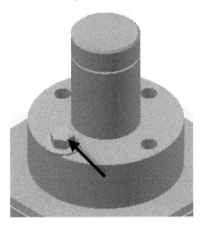

Patterning components in an assembly

1. On the ribbon, click **Assemble > Pattern > Pattern**.

2. Select the **Bolt connection** from the Browser window.

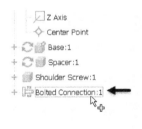

3. On the **Pattern Component** dialog, click the **Circular** tab and select the **Axis Direction** button.

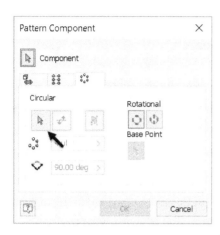

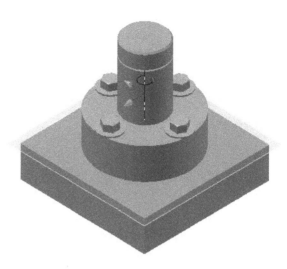

4. Click on the large cylindrical face of the Spacer to define the axis of the circular pattern.
5. On the dialog, type in 4 and 90 in the **Circular Count** and **Circular Angle** boxes, respectively.
6. Click **OK** to pattern the bolt connection.

2. On the ribbon, click **Assemble > Relationships > Constrain**.
3. On the dialog, click the **Mate** icon and click on the round faces of the Spacer and Base.

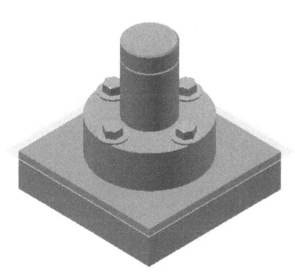

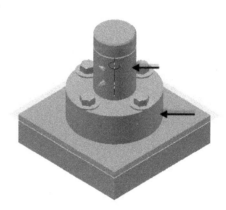

Applying the constraint to the components

1. On the ribbon, click **View > Visibility > Degrees of Freedom**.

4. Click the **Aligned** icon in the **Solution** group.
5. Click **Apply**.
6. Click on the round faces of the Shoulder Screw and Base.

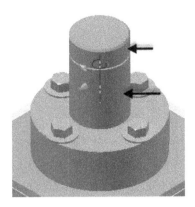

7. Click **Apply**.
8. On the dialog, select **Flush** from the **Solution** section.
9. In the Browser Window, expand the **Origin** folder and select XY Plane.

10. Expand the **Origin** folder of the Shoulder Screw and select XZ Plane.

11. Click **OK** to fully-constrain the assembly.
12. Save the assembly and all its parts.

Using the Search tool in the Browser window

Autodesk Inventor 2020 provides you with the search tool to locate the components or features very quickly.

1. In the Browser Window, click the **Search** icon.
2. Type 'hex' in the search bar; all the hexagonal bolts appear in the browser window.

3. Place the pointer on the hexagonal bolts in the browser window; they are highlighted in the graphics window.

You can select all the hexagonal bolts by pressing the Shift key and clicking on them. After selecting them, you can perform a variety of operations at a time such as hiding, deleting, solving, suppressing, and so on.

4. Click **Clear Search** to clear all the searched components.

Editing Values in the Browser window

Autodesk Inventor 2020 allows you to edit the values of the assembly components directly in the Browser Window.

1. In the Browser Window, click the Drop-down menu next to the Search box, and then select **Edit Values in Browser**.

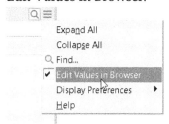

2. In the Browser Window, expand the **Shoulder Screw** part, and then click on the **Flush** relation, as shown; the selected relation is highlighted in the graphics window, as shown.

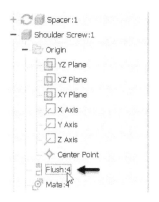

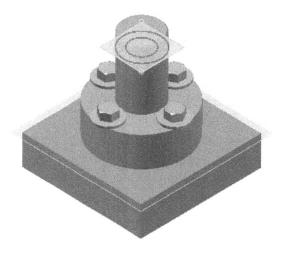

3. Type 1 in the box that appears next to the selected relation, and then press Enter; the relation is updated in the graphics window.

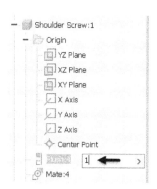

4. Click **Undo** on the Quick Access Toolbar.

Changing the Display Preferences of the Browser window

Autodesk Inventor allows you to hide or display items to reduce the clutter in your Browser Window. For example, you can hide or display the work features such as plane and UCS in the Browser Window.

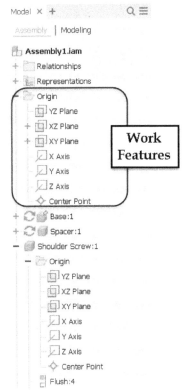

1. In the Browser Window, click the Drop-down menu next to the **Search** box, and then select **Display Preferences > Hide Work Features**.

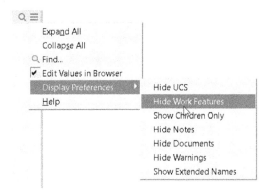

The work features are hidden.

Using the Measure tool

The **Measure** tool helps you measure the size and position of the model. You can measure the various parameters of the model, such as length, angle, radius, and so on.

1. On the ribbon, click **Inspect** tab > **Measure** panel > **Measure** ; the **Measure** floating window appears on the screen.

2. Click and drag the **Measure** floating window, and then release it on to the Browser Window; the Measure window is docked to the Browser Window.

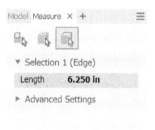

The length of the selected edge is displayed in the **Measure** window.

The **Measure** Window has three selection filters (shown from right to left): **Select faces and edges**, **Part Priority**, and **Component Priority**.

The **Select faces and edges** filter allows you to select only the faces and edges of the model.

The **Part Priority** filter allows you to select the part geometry for measurement.

The **Component Priority** filter allows you to select the part geometry and assemblies. This filter is used to select subassemblies from the main assembly.

3.   Select the **Select faces and edges** filter and select the straight edge, as shown.

4.   Click **Advanced Settings** in the **Measure** window.

In the **Advanced Settings** section, you can change the **Precision**, **Angle Precision** of the displayed measurement. In addition to that, you can display the measurement in dual units by specifying the **Dual Units** type.

5.   Select the round face, as shown; the Measure window displays results.

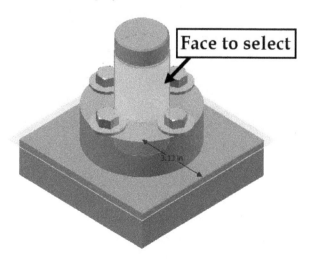

The Measure results section display the results of

the first and second selections separately. In addition to that, the distance between the two selected entities is displayed.

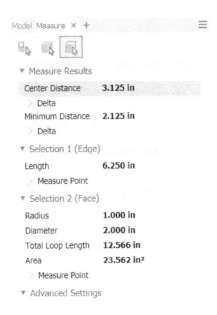

6. Save and close the assembly and its parts.

TUTORIAL 2

In this tutorial, you create a slider crank mechanism by applying Joints.

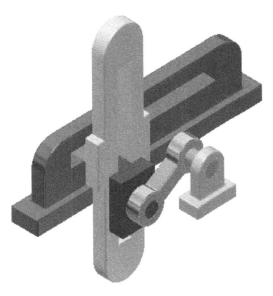

1. Create the **Slider Crank Assembly** folder inside the project folder.
2. Download the part files of the assembly from the companion website. Next, save the files in

the **Slider Crank Assembly** folder.

3. Start a new assembly file using the **Standard.iam** template.
4. Click **Assemble > Component > Place** on the ribbon.
5. Browse to the **Slider Crank Assembly** folder and double-click on **Base**.
6. Right-click and select **Place Grounded at Origin**.
7. Right click and select **OK**.

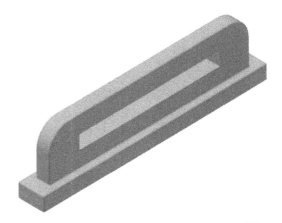

8. Click **Assemble > Component > Place** on the ribbon.
9. Browse to the **Slider Crank Assembly** folder and select all the parts except the **Base**.
10. Click **Open** and click in the graphics window to place the parts.
11. Right click and select **OK**.
12. Click and drag the parts if they are coinciding with each other.

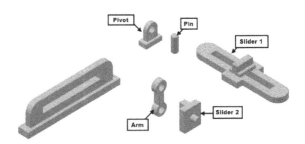

Creating the Slider Joint

1. Click **Assemble > Relationships > Joint** on the

ribbon; the **Place Joint** dialog appears.

2. Set the **Type** to **Slider**.

3. Select the face on the Slider1, as shown below.

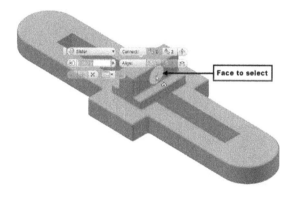

4. Select the face on the Base, as shown below; the two faces are aligned.

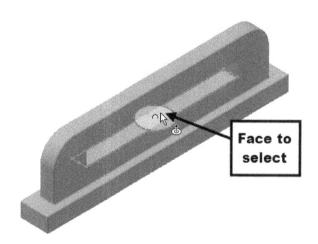

5. On the dialog, click the **First Alignment** button.

6. Select the face of the Slider1, as shown.

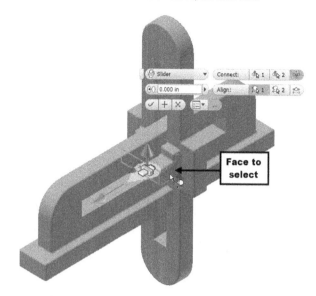

7. Select the face of the Base, as shown; the Slider 1 is aligned to the selected face.

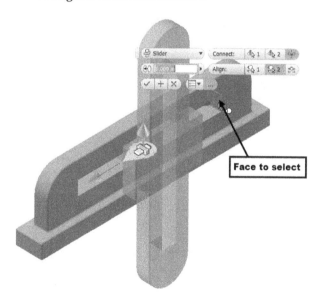

5. Click the **Limits** tab on the **Place Joint** dialog.
6. Check the **Start** and **End** options under the **Linear** group.
7. Set the **Start** value to 1.5 in and **End** to -1.5 in.

8. Click **OK**.
9. Select the Slider1 and drag the pointer; the Slider1 slides in the slot of the Base. Also, the slider motion is limited up to the end of the slot.

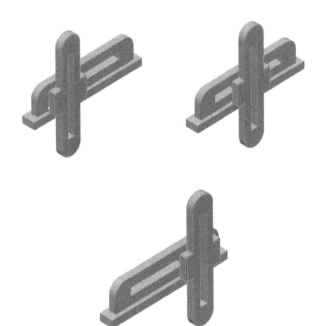

10. Click the corner of the ViewCube, as shown; the orientation of the assembly is changed.

11. Click **Assemble > Relationships > Joint** on the ribbon.

12. On the dialog, set the **Type** to **Slider**.
13. Select the face on the Slider2, as shown below.

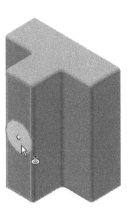

14. Select the right edge of the top face of the ViewCube; the orientation of the assembly changes.

15. Select the face on the Slider1, as shown below. Next, click **OK**.

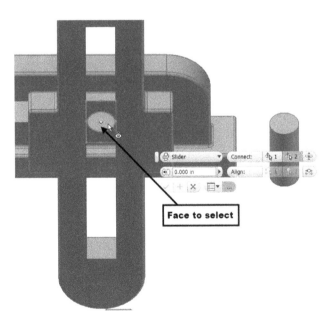

Face to select

Creating the Rotational Joint

1. Click **Assemble > Relationships > Joint** on the ribbon.
2. Set **Type** to **Rotational**.

Type

3. Select the circular edge of the arm, as shown below.

4. Select the circular edge of the Slider2.

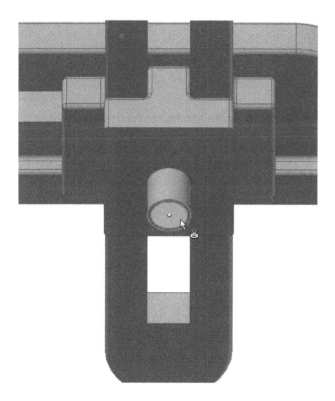

6. Click the **Flip Component** button under the

Connect group.

7. Click **OK**

Creating the Rigid Joint

1. Click **Assemble > Relationships > Joint** on the ribbon.

2. Set the **Type** to **Rigid**.

Type

3. Select the top face on the pin.

4. Click on the corner point of the ViewCube, as shown.

5. Select the circular edge on the back face of the arm.

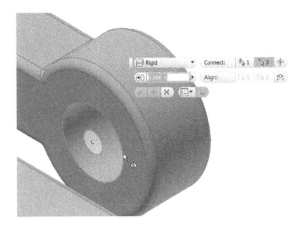

6. Click **OK**.

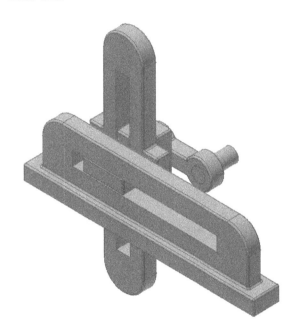

Adding more assembly joints

1. Create another rotational joint between the Pin and the Pivot.

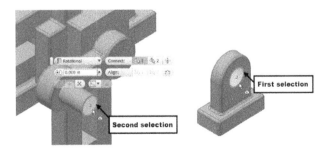

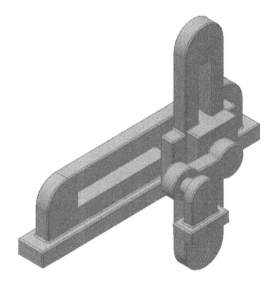

Next, you need to constrain the Pivot by applying constraints.

2. Click the **Assemble** button on the **Relationships** panel.

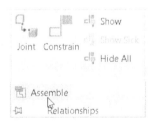

10. On the **Assembly** mini toolbar, select **Mate – Flush** from the drop-down.

11. Select the bottom face of the Pivot, and then select the bottom face of the Base.

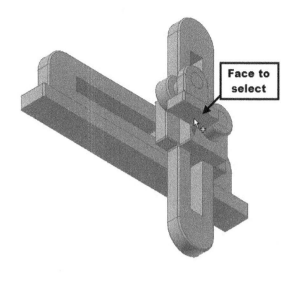

14. Click **OK** (checkmark on the mini toolbar).

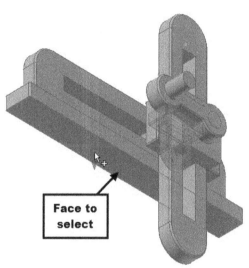

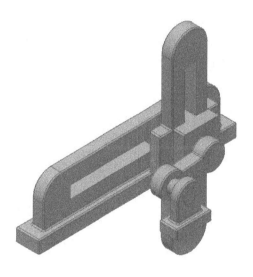

12. Click **Apply** (plus symbol on the mini toolbar).
13. Select the **XY Plane** of the Pivot and **XY Plane** of the Base from the **Browser window**.

Driving the joints

1. In the Browser window, expand Pivot and click the right mouse button on the **Rotational** joint.
2. Select **Drive** from the shortcut menu.

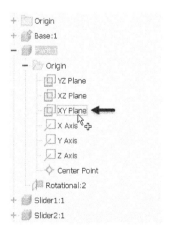

189

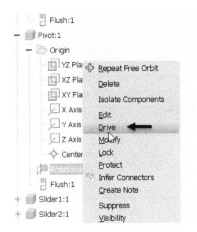

3. On the **Drive** dialog, type in 0 and 360 in the **Start** and **End** boxes, respectively.

4. Expand the dialog by clicking the double-arrow button located at the bottom. On the expanded dialog, you can define the settings such as drive adaptivity, collision detection, increment, repetition, and so on.

5. Click the **Record** ⊙ button on the dialog. Specify the name and location of the video file — Click **Save** and **OK**.

6. On the dialog, click the **Forward** ▶ button to simulate the motion of the slider-crank assembly.

7. Click **OK** to close the dialog.

Creating Positions

1. In the Browser window, expand the **Representations > View** and notice that the **Master** representation is set as default.

2. Right click on the **Position** node, and then select **New**; a new position is created.

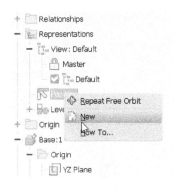

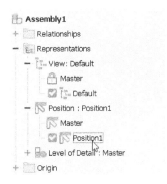

3. Double click on the **Position1** and type **StartPosition**; the view representation is renamed.

4. Click and drag the Slider1 to the left end, as shown.

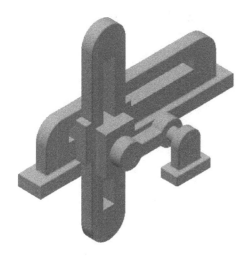

5. Double click on the **Master** position to activate it.

Creating 3D PDF

Autodesk Inventor allows you to create a 3D PDF from the model. The 3D PDF file is helpful in viewing the 3D without any CAD application or viewer.

1. Click **File Menu > Export > 3D PDF**.

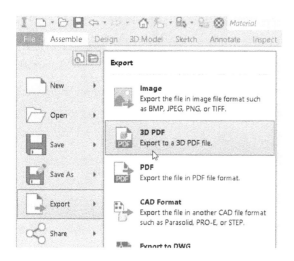

The **Publish 3D PDF** dialog appears on the screen. It is powered by **Anark Core** software. On the Publish 3D PDF dialog, you can select the properties to be displayed on the PDF from the **Properties** section. You can also select the required design view representation, visualization quality, and export scope.

2. Leave the Template to the default setting.

If required, you can select another template by clicking the icon next to the Template path. You can also create a new 3D PDF template if you have Adobe Acrobat Pro. You can go through the Autodesk Inventor Help file to know the procedure to create a 3D PDF template.

3. Specify the **File Output Location**.
4. Check the **View PDF when finished** option.
5. Check the **Generate and attach STEP file** option.
6. Click the **Options** button next to the **Generate and attach STEP file** option; the **STEP file Save as Options** dialog appears on the screen.

On this dialog, select the required Application Protocol option and spline fit accuracy. You can also enter the authorization, author, organization, and description.

7. Click **OK** on the **STEP file save as Options** dialog.

Use the **Attachments** button, if you want to add any other attachments to the PDF file such as a spreadsheet, pdf, or text document.

8. Click **Publish** on the **Publish 3D PDF** dialog.

Inventor starts exporting the 3D model to the PDF file. After a few seconds, the 3D PDF file opens in the PDF viewer.

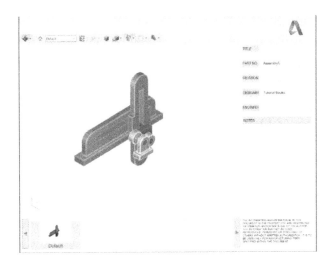

9. Click inside the graphics window of the PDF file, and then drag to rotate the model.

10. Click the drop-down located at the top left corner and notice the **View** options. These options are the same as that available on the Navigation Pane.

You can open or save the STEP file by right-clicking on it and select the corresponding option.

12. Close the 3D PDF file
13. Save and close the assembly and its parts.

Likewise, examine the other options on the toolbar. These options are similar to that available on the View ribbon tab of the Autodesk Inventor application.

11. On the Sidebar, click the Attachments icon to view the STEP file.

Chapter 9: Dimensions and Annotations

In this chapter, you will learn to

- Create Centerlines and Centered Pattern
- Edit Hatch Pattern
- Apply Dimensions
- Place Hole callouts
- Place Leader Text
- Place Datum Feature
- Place Feature control frame
- Place Surface texture symbol
- Modify Title Block Information

TUTORIAL 1

In this tutorial, you create the drawing shown below.

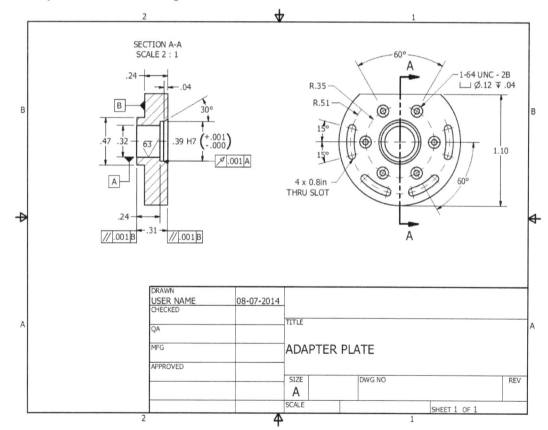

1. Open a new drawing file using the **Standard.idw** template.
2. In the Browser Window, click the right mouse on Sheet:1 and select **Edit Sheet**.
3. On the **Edit Sheet** dialog, select **Size > A**, and then click **OK**.
4. Click **Place Views > Create > Base** on the ribbon.
5. Click **Open an existing file** button on the dialog.
6. Browse to the location of the Adapter Plate created in Tutorial 1 of Chapter 5. You can also download this file from the companion website and use it.
7. Select the Adapter Plate file and click **Open**.
8. Set the **Scale** to 2:1.
9. Click the Front face on the ViewCube displayed in the drawing sheet.
10. Set the **Style** to **Hidden Line Removed**.
11. Click **OK** on the dialog.
12. Drag the view to the right side of the drawing sheet.
13. Click **Place Views > Create > Section** on the ribbon.
14. Select the front view.
15. Draw the section line on the front view.

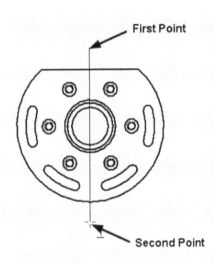

15. Right-click and select **Continue**.
16. Place the section view on the left side.

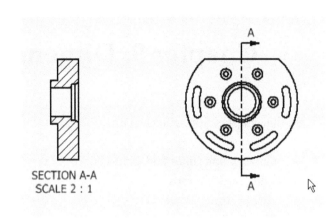

SECTION A-A
SCALE 2 : 1

Creating Centerlines and Centered Patterns

1. Click **Annotate > Symbols > Centerline Bisector** on the ribbon.

2. Select the parallel lines on the section view, as shown below; the centerline is created.

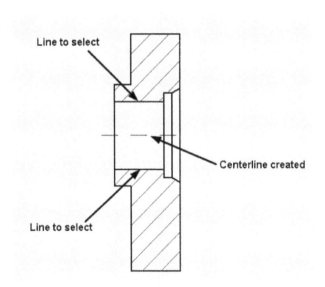

3. Click **Annotate > Symbols > Centered Pattern** on the ribbon.

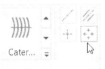

4. Select the circle located at the center.

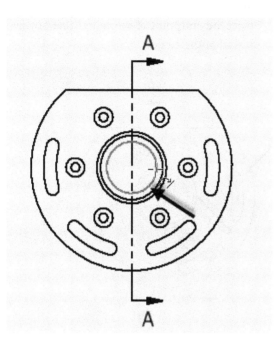

5. Select the center point of any one of the counterbored holes.

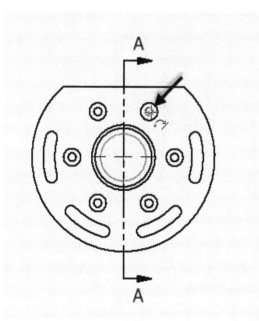

6. Select the center points of other counterbored holes.
7. Click the right mouse button and select **Create**.

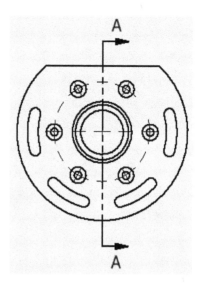

8. Likewise, create another centered pattern on the curved slots: Right-Click and select **Create**.

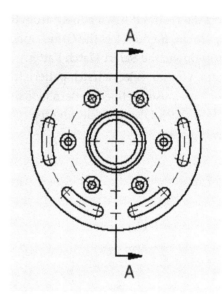

9. Press Esc to deactivate the tool.

Editing the Hatch Pattern

1. Double-click on the hatch pattern of the section view; the **Edit Hatch Pattern** dialog appears.

You can select the required hatch pattern from the **Pattern** drop-down. If you select the **Other** option from this drop-down, the **Select Hatch Pattern** dialog appears. You can select a hatch pattern from this dialog or load a user-defined pattern by using the **Load** option. Click **OK** after selecting the required hatch pattern.

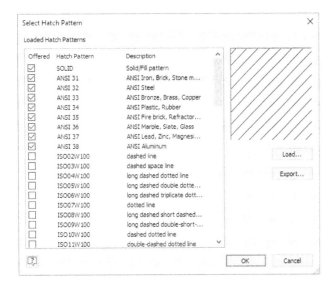

2. Click **OK**.

Applying Dimensions

1. Click **Annotate > Dimension > Dimension** on the ribbon.

2. Select the center line on the slot located at the left.
3. Select the endpoint of the center line of the hole located at the center.

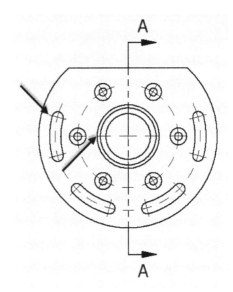

4. Move the pointer toward left and click.
5. Click **OK**.

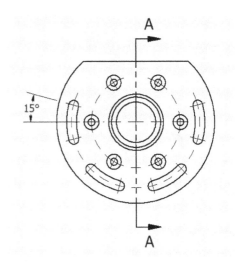

6. Likewise, create another angular dimension, as shown below.

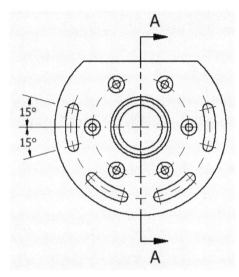

7. Create angular dimensions between the holes, and then between slots. To create the angular dimension between the slot, you need to create the angular dimension between the bolt circles.

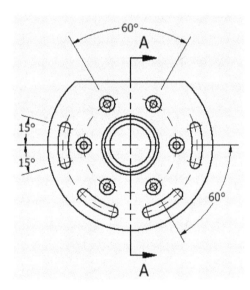

8. Dimension the pitch circle radius of the slots (activate the **Dimension** command and select th centreline passing through the slots. Next, right click and select **Dimension type > Radius**).

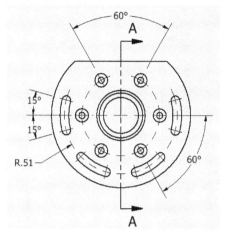

9. With the **Dimension** tool active, select the horizontal line of the front view and the lower quadrant point of view.

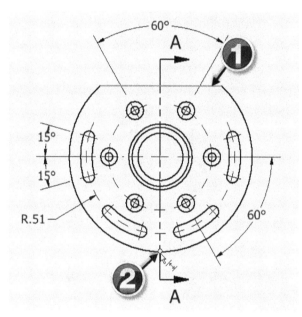

10. Place the dimension on the right side. Click **OK**.

197

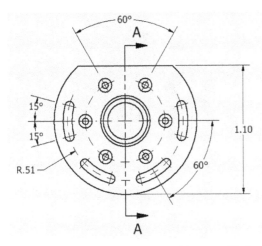

11. Click **Annotate > Feature Notes > Hole and Thread** on the ribbon.

12. Select the counterbore hole and place the hole callout, as shown below.

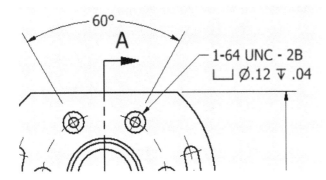

13. Add a pitch circle radius to counterbore holes.

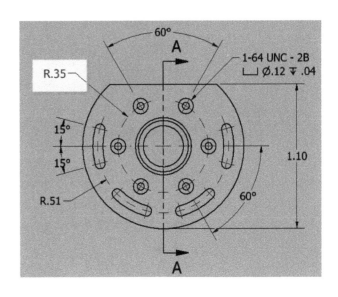

14. Click **Leader Text** on the **Text** panel.

15. Select the slot end, as shown below.

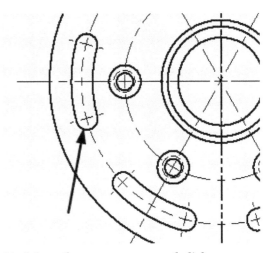

16. Move the cursor away and click.
17. Right-click and select **Continue**; the **Format Text** dialog appears.
18. Enter the text shown below.

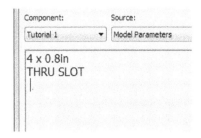

19. Click **OK** — Press the Esc key.

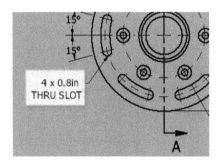

20. Double-click on the section label below the section view.
21. On the **Format Text** dialog, select all the text and set the **Size** to **0.120**. Click **OK**.

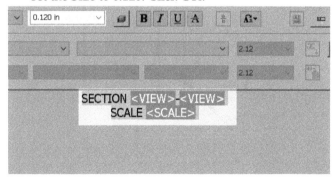

22. Drag and place the section label on the top.

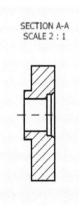

23. Click **Dimension** on the **Dimension** panel.
24. Select the lines, as shown below.

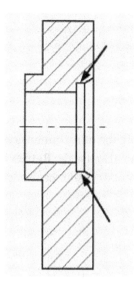

25. Move the pointer toward the right and click to place the dimension.
26. On the dialog, click the **Precision and Tolerance** tab.
27. Set the **Tolerance Method** to **Limits/Fits - Show tolerance**.
28. Select **Hole > H7**.
29. Set the **Primary Unit** value to **3.123**.
30. Set the **Primary Tolerance** value to **3.123**.

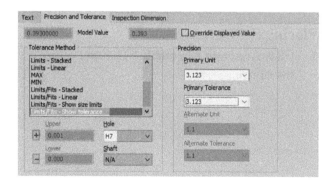

31. Click **OK**.

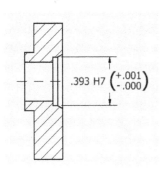

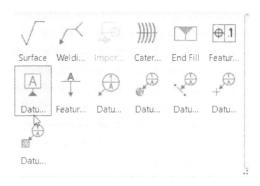

32. Likewise, apply the other dimensions, as shown below. You can also use the **Retrieve Dimensions** tool to create the dimensions.

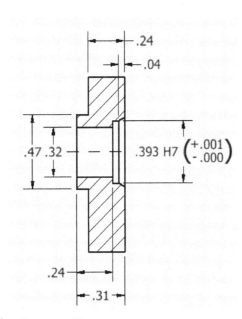

Placing the Datum Feature

1. Click **Annotate > Symbols > Datum Feature** on the ribbon.

2. Select the extension line of the dimension, as shown below.

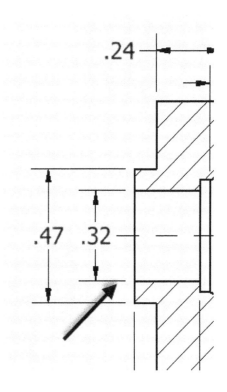

3. Move the cursor downward and click.
4. Move the cursor toward left and click; the **Format Text** dialog appears. Make sure that A is entered in the dialog.
5. Click **OK**.

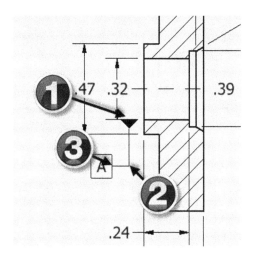

6. Likewise, place a datum feature B, as shown below. Press Esc.

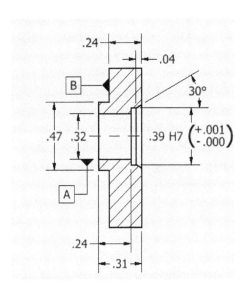

Placing the Feature Control Frame

1. Click **Annotate > Symbols > Feature Control Frame** on the ribbon.

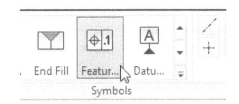

2. Select a point on the line, as shown below.
3. Move the cursor horizontally toward the right and click.

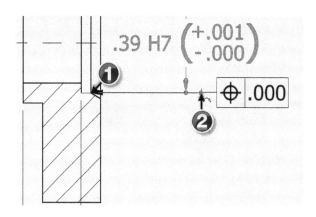

4. Right-click and select **Continue**; the **Feature Control Frame** dialog appears.

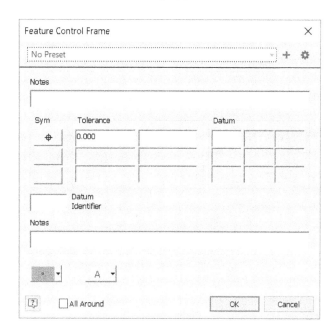

5. On the dialog, click the **Sym** button and select **Circular Run-out**.

6. Enter 0.001 in the **Tolerance** box and **A** in the **Datum** box.

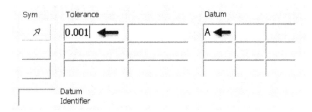

7. Click **OK**.

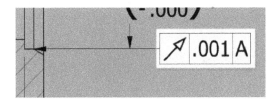

8. Right-click and select **Cancel**.

Placing the Surface Texture Symbols

1. Click **Annotate > Symbols > Surface Texture Symbol** on the ribbon.

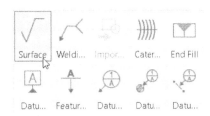

2. Click on the inner cylindrical face of the hole, as shown below.

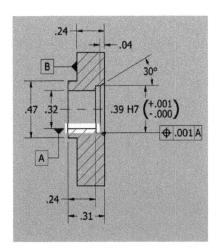

3. Right-click and select **Continue**; the **Surface Texture** dialog appears.

4. Set the **Roughness Average - maximum** value to 63.

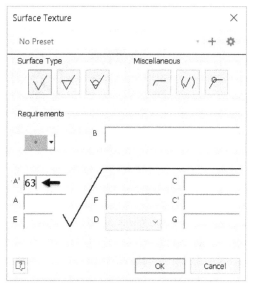

5. Click **OK**.

6. Right-click and select **Cancel**.

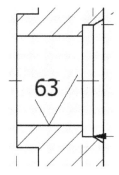

7. Apply the other annotations of the drawing. The final drawing is shown below.

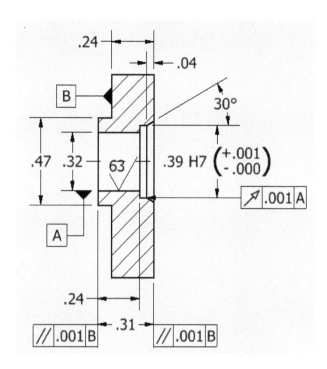

respective tabs.

3. Click **OK**.

4. Save the file.

5. To export the file to AutoCAD format, click **File Menu > Export > Export to DWG**.

6. Click **Save**.

7. Close the file.

Modifying the Title Block Information

1. Right-click on the **Adapter Plate** in the **Browser window**. Select **iProperties** from the shortcut menu.

2. Click the **Summary** tab and enter the information, as shown next.

You can also update the Project information, drawing status, and other custom information in the

Dimensions and Annotations

Chapter 10: Model-Based Dimensioning

Geometric Dimensioning and Tolerancing

During the manufacturing process, the accuracy of a part is an essential factor. However, it is impossible to manufacture a part with the exact dimensions. Therefore, while applying dimensions to a drawing you need to provide some dimensional tolerances, which lie within acceptable limits. The following figure shows an example of dimensional tolerances applied to the drawing.

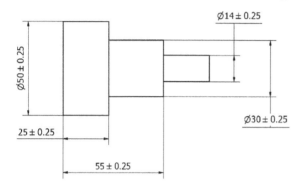

The dimensional tolerances help you to manufacture the component within a specific size range. However, the dimensional tolerances are not sufficient for manufacturing a component. You must give tolerance values to its shape, orientation, and position as well. The following figure shows a note, which is used to explain the tolerance value given to the shape of the object.

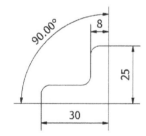

Providing a note in a drawing may be confusing. To avoid this, we use Geometric Dimensioning and Tolerancing (GD&T) symbols to specify the tolerance values to shape, orientation, and position of a component. The following figure shows the same example represented by using the GD&T symbols. In this figure, the vertical face to which the tolerance frame is connected must be within two parallel planes 0.08 apart and perpendicular to the datum reference (horizontal plane).

Model-Based Dimensioning

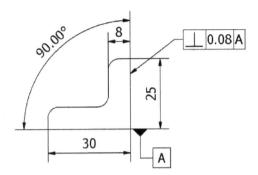

Providing GD&T in 2D drawings is a common and well-known method. However, you can provide GD&T information to 3D models as well. The tools available in the **Annotate** tab of the ribbon help you to add GD&T information to the 3D models based on the universal standards such as ASME Y14.41 – 2003 and ISO 16792: 2006. However, you can add GD&T information based on your custom standard as well.

In this chapter, you will learn to use **Annotate** tools to add GD&T information to the part models. There are many ways to add GD&T information and full-define the parts and assemblies. There are few methods explained in this chapter, but you need to use a method, which is most suitable for your design.

TUTORIAL 1

This tutorial teaches you to add tolerances to the 3D Model.

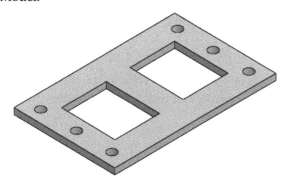

1. Download the Model-Based Dimensioning part files from the Companion website and open the Tutorial 1 file.
2. On the ribbon, click the **Tools** tab > **Options** panel > **Document Settings** to open the **Document Settings** dialog.
3. Click the **Standard** tab and select **ASME** from the **Active Standard** drop-down.
4. Click **OK**.
5. In the Browser Window, expand the **View** node, and then double click on the **Isometric** view.
6. Right click on the **Isometric** view, and then select **Annotation Scale > Auto**.

You can also change the **Annotation Scale** from the **Annotation Scale** drop-down available on the **Manage** panel of the **Annotate** ribbon tab.

Adding Tolerances to the Model

1. On the ribbon, click **Annotate** tab > **Geometric Annotation** panel > **Tolerance Advisor** ; the **Tolerance Advisor** panel appears.
2. Click the **Face status Coloring** option located at the bottom on the **Tolerance Advisor** panel.

3. On the ribbon, click **Annotate** tab > **Geometric Annotation** panel > **Tolerance Feature** .
4. Click on the top face of the model.
5. Click on the bottom face of the model.

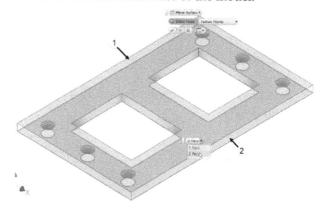

6. Select **Slab** from drop-down located on the Mini toolbar.

7. Click **OK**.
8. Right click and select the **Select Annotation Plane [Shift]** option.

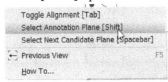

9. Select the face of the model, as shown.

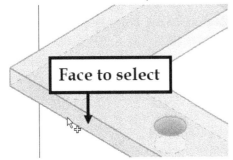

10. Move the pointer toward left, and then click.
11. Click on the dimension value.

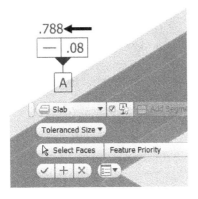

12. Select **Symmetric** from the **Tolerance Type** drop-down.

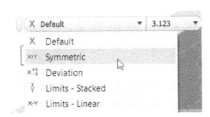

13. Click on the tolerance value displayed next to

the dimension value.
14. Type .002 in the **Tolerance** box.

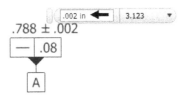

15. Click on the tolerance value in the feature control frame, and then type .002 in the **Tolerance** box.

16. Click **OK**; top and bottom faces of the model are displayed in green, which means that they are fully-constrained.

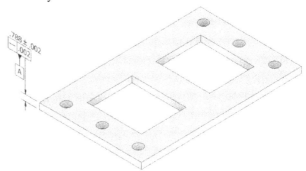

17. On the ribbon, click **Annotate** tab > **Geometric Annotation** panel > **Tolerance Feature** .
18. Select the **Slab** option from the drop-down.

19. Click **OK**.
20. Press and hold the Shift key, and then select the top face of the model.
21. Move the pointer toward the right and click to place the tolerance feature.

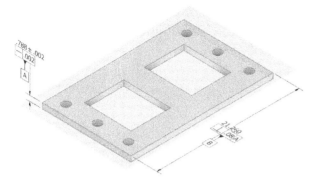

22. Specify the tolerance values, as shown.

23. Click **OK** on the Mini toolbar.
24. Likewise, add the tolerance feature between the left and right side faces, as shown.

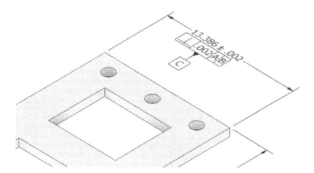

25. Activate the **Tolerance Feature** command.
26. Select the faces of the cut feature, as shown.

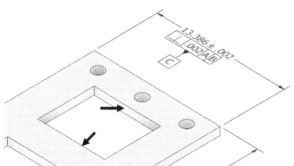

27. Select the **Slot** option from the drop-down from the Mini toolbar.

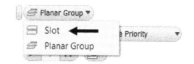

28. Click **OK**.
29. Press and hold the Shift key, and then select the top face of the model.
30. Move the pointer toward the left and click to place the tolerance feature.
31. Specify the tolerance values, as shown.

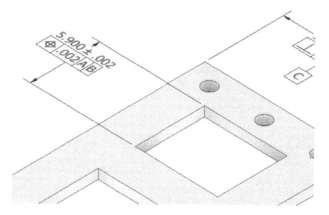

32. Likewise, add other tolerance features, as shown.

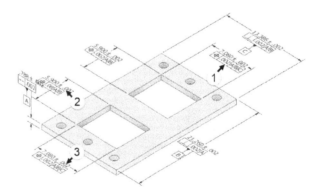

33. Activate the **Tolerance Feature** command.
34. Select any one of the holes.
35. Select the **Select Hole Parallel Axes Pattern** option from the Mini toolbar.
36. Click **OK**.
37. Press and hold the Shift key, and then select the top face of the model.
38. Place the tolerance feature and add the tolerance values.

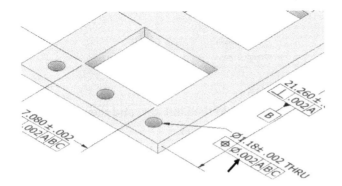

Notice that all the elements of the model are highlighted in green, which means that they are fully constrained.